INTUITION IN SCIENCE AND MATHEMATICS

MATHEMATICS EDUCATION LIBRARY

EFRAIM FISCHBEIN
School of Education, Tel Aviv University, Israel

INTUITION IN SCIENCE AND MATHEMATICS

An Educational Approach

SPRINGER-SCIENCE+BUSINESS MEDIA, B.V.

Library of Congress Cataloging-in-Publication Data

CIP

Fischbein, Efraim, 1920–
Intuition in science and mathematics: an educational approach / Efraim Fischbein.
p. cm. — (Mathematics education library)
Bibliography: p.
Includes index.

DOI 10.1007/978-0-306-47237-4
1. Science—Philosophy. 2. Mathematics—Philosophy. 3. Intuition. I. Title. II. Series.
Q175.F5468 1987 87–26203
501—dc 19

02-0194-200 ts
Second printing 1994

TABLE OF CONTENTS

PART II: FACTORS WHICH SHAPE INTUITIONS

PREFACE

In writing the present book I have had in mind the following objectives:

— To propose a theoretical, comprehensive view of the domain of intuition.
— To identify and organize the experimental findings related to intuition scattered in a wide variety of research contexts.
— To reveal the educational implications of the idea, developed for science and mathematics education.

Most of the existing monographs in the field of intuition are mainly concerned with theoretical debates — definitions, philosophical attitudes, historical considerations. (See, especially the works of Wild (1938), of Bunge (1962) and of Noddings and Shore (1984).)

A notable exception is the book by Westcott (1968), which combines theoretical analyses with the author's own experimental studies.

But, so far, no attempt has been made to identify systematically those findings, spread throughout the research literature, which could contribute to the deciphering of the mechanisms of intuition. Very often the relevant studies do not refer explicitly to intuition. Even when this term is used it occurs, usually, as a self-evident, common sense term.

The explanation is that intuition is generally seen as a primary phenomenon which may be described but which is not reducible to more elementary components. As a matter of fact, *intuitively,* intuition has the appearance of a self-evident, self-consistent cognition, like the perception of a color or the experience of an emotion. The effect is that, generally, no attempt is made by researchers to use their experimental findings for elucidating the structure of intuitive phenomena. On the contrary: it is intuition which is used as a descriptive and explanatory concept. Piaget, for instance, who used the terms intuition and intuitive thinking very often never made any attempt to refer to his own findings for obtaining a deeper understanding of the general structure of intuitive mechanisms as such. With a very few exceptions this is the situation with most of the researchers. Such an exception is the work of Andrea DiSessa (1982, 1983a, 1983b) who made an important attempt to analyse his own experimental findings with the explicit purpose of building a theory of intuition.

The present book has a similar aim. We expound a theory, we try to get a richer insight in the mechanisms of intuition using research evidence, and we try to support and enlarge our conception on the basis of these data.

Our general thesis is that, in principle, cognition fulfils behavioral aims and is shaped by behavioral constraints. The same is to be said about intuition which is a particular form of cognition.

Intuition has its roots in the syncretic type of thinking of the child and of human beings in the early stages of civilization. But it does not survive in adults and in highly developed cultures only as a mere residuum. We claim that *intuition expresses a profound necessity of our mental behavior.*

During the very course of our reasoning, of our trial-and-error attempts, we have to rely on representations and ideas which appear, subjectively, as certain, self-consistent and intrinsically clear. We cannot doubt everything at every moment. This would be a paralysing attitude. Some representations, some conceptions have to be taken for granted. They have to appear, subjectively, as autonomous, coherent, totally and directly acceptable cognitions in order to keep the process of reasoning working fruitfully.

An intuition is, then, such a crystallized — very often prematurely closed — conception in which incompleteness or vagueness of information is masked by special mechanisms for producing the feelings of immediacy, coherence and confidence.

Such mechanisms have been described in the research literature, but very often without any apparent connection with a theory of intuition. In the present work an attempt has been made to take advantage of these research sources. Studies in overconfidence, in subjective probabilities, findings referring to mental models, to typical errors in naive physics, to misconceptions in mathematics, to the evolution of logical concepts in children etc. represent, in fact, rich potential sources for a theory of intuition.

I should like to emphasize again this interesting phenomenon: simply because intuition is tacitly but firmly considered to be a primitive feeling, rich sources of information, based on experimental findings, have been ignored by most of the theorists. A primary purpose of this book has been to overcome this obstacle.

The history of science and mathematics is also an important source for understanding the dramatic struggle of the scientific mind against intuitive biases. We have used a number of examples in our work in this respect, trying to identify the common structure of intuitive difficulties in experts and novices.

Turning our attention, now, to the educational aspects: Most authors — experimental researchers and theorists alike — strive to set up recommendations for avoiding intuitively-based errors in learning and problem solving and for improving intuitive guesses and evaluations. But, in our opinion, one can hardly expect such suggestions to be really helpful if they are not based on a comprehensive theory of intuition. Intuitions are only apparently autonomous, self-evident cognitions. They are so in order to confer on some of the individual's ideas the appearance of certitude and intrinsic validity. But, in fact, these ideas appear very robust as an effect of their being deeply rooted in the person's basic mental organization. Consequently, in order to eliminate or change or even control an intuitive attitude it would be necessary to produce a profound, structural transformation in large areas of mental activity.

Therefore, the individual's self-confidence itself might be endangered if he learns that even his deepest beliefs may very often be misleading. It follows that intuitions cannot be treated effectively and positively as mere isolated symptoms but rather as manifestations of highly articulated and very complex structures. And for this, a comprehensive theory is necessary, a theory which would take into acount the behavioral roots and the adaptive functions of intuitions.

I do not pretend to have solved all these puzzling theoretical problems. It is probably too early to get definite answers. But, if I have been able to convince the reader of the importance of this line of thought, if I have been able to cast some doubts and to produce some new expectations with regard to his intuition *about* intuition then my main goal will have been achieved.

The book is divided into two major parts. The first part is concerned with the theoretical aspects — the theory, the relevance of intuitive forms of cognition for scientific and mathematical reasoning, the connections between intuition and other, related, categories of cognition, the general characteristics and the classification of intuitions.

The second part of the book deals mainly with factors which contribute to shaping intuitions: the role of experience, the role of various types of models — analogies, paradigms, diagrams, phenomenological primitives — the role of factors for producing the effects of immediacy and globality.

I believe that the time is now ripe for a formal recognition of intuition as one of the major components of our cognitive endeavors. A constant interplay between theory and experiment in this field is nowadays certainly possible and highly desirable.

It is our conviction that the findings will be beneficial for both cognitive psychology and educational practice.

ACKNOWLEDGMENTS

For the past ten years I have given lectures and directed seminars on intuition at the School of Education of Tel Aviv University. This book is largely based on that activity. My gratitude goes, then, first to my students and those colleagues who patiently helped me during all these years to clarify my ideas and to see new problems where everything seemed to be settled.

I am very grateful to Alan Bishop for inviting me to write this book and for encouraging me to continue my work whenever he felt that the rhythm was slowing down.

With my son-in-law Dr. Amir Klein, a brilliant physicist, I had numerous discussions about the role of intuition in scientific creativity. I thank him deeply for his generous and kind collaboration.

I am also grateful to Brian Greer for his constructive contribution to the work of preparing the final manuscript, and extend my thanks to the staff of the D. Reidel Publishing Company for their constant and efficient cooperation.

Last, but by no mean least, my deepest gratitude goes to my wife Henrietta who, with intelligence, devotion and competence, revised every sentence and helped me to overcome the unpleasant moments of fatigue and uncertainty.

PART I

THE THEORY

CHAPTER 1

INTUITION AND THE NEED FOR CERTITUDE

THE COMPLEXITY OF THE DOMAIN

Intuition is certainly a highly controversial concept in science and philosophy. Accepted by some as the basic source of every true knowledge, rejected by others as potentially misleading every quest for truth, intuition — as a concept and as a method — revives itself again and again in philosophical disputes, in the theoretical foundations of science and mathematics, in mystical considerations, in ethics and aesthetics, in pedagogy, and yet very little and very seldom in psychology.

The Variety of Meanings

In some contexts, intuition is referred to as a *source* of true — or apparently true — knowledge. It is generally in this sense that the term *intuition* is used in the works of Descartes (1967) and Spinoza (1967). For both of them, in a world of misleading appearances and futile interpretations, intuition remains the ultimate reliable source of absolutely certain truths.

For others, intuition is rather *a method*, a sort of mental strategy which is able to reach the essence of phenomena. Bergson has been the main advocate of this usage. According to Bergson, intelligence addresses itself to the world of objects, of solids, of static realities. In order to understand reality, intelligence uses a "cinematographic" procedure: the uninterrupted flow of real phenomena is cut into sequences of static representations mainly expressed in concepts. But the essence of motion, of life, of spirit, of duration cannot be reached this way. According to Bergson it is through intuition — a kind of sympathetic identification — that we are able to grasp the very essence of living and changing phenomena (Bergson, 1954).

The term intuition is also used for indicating a certain category of cognitions, i.e. cognitions which are directly grasped without, or prior to, any need for explicit justification or interpretation. It is in this sense that Piaget refers to spatial and temporal intuitions, to empirical and operational intuitions, to pure intuitions etc. (Beth and Piaget, 1961).

In Kant's terminology the concept of intuition gets a more restrictive meaning, compared with those referred to in the previous presentations. Namely, according to Kant, intuition is simply the faculty through which objects are directly grasped in distinction to the faculty of understanding through which we achieve conceptual knowledge. Kant uses the terms *intellectual* and *sensible* intuitions, but practically, it is only the sensible variant which makes sense to him. An "intellectual intuition" would be necessary for knowing the "noumenon", the reality in itself — and this is impossible. Therefore in Kant's terminology intuition remains related to sensorial knowledge, while an "intellectual" intuition simply does not exist (Kant, 1980, p. 268).

The term "intellectual intuition" may also be used for designating forms of immediate knowledge which are not sensorial, which deal with concepts, formal relations, theories. One may affirm, for instance, that the statement: "Every natural number has a successor" is intuitively acceptable, and in this case we have an "intellectual" intuition. In contrast, the intuitive evaluation of the weight of an object or of the speed of a moving body would represent sensorial intuitions. Certainly, no clear-cut distinction is possible, but the terminology related to intuition is so confusing that we feel that even in these introductory lines the reader should get a preliminary picture of the complexity of this domain.

As has already been noticed with regard to the *credibility* of intuitions, we may also identify various, even opposite, conceptions. Sometimes, intuition is referred to as a global guess for which an individual is not able to offer a clear and complete justification. Very often *intuition* means an elementary, common sense, popular, primitive form of knowledge, as opposed to scientific conceptions and interpretations. In contrast, according to some philosophers, like Spinoza, intuition is the highest form of knowledge through which the very essence of things, and God Himself, may be revealed. According to Poincaré, no genuine creative activity is possible in science and in mathematics without intuition, while for Hahn (1956) intuition is mainly a source of misconceptions and should be eliminated from a serious scientific endeavor.

In the pedagogical literature, intuition is often related to sensorial knowledge as the first necessary basis for a further intellectual education. In this sense, intuitive knowledge is, more or less, equivalent to perceptual knowledge (i.e. concrete objects, pictures, diagrams). In some educational approaches one advocates the necessity to use a large amount of intuitive (concrete, pictorial, manipulative) devices, while according to others one has to eliminate, as soon and as far as possible, intuitive techniques, especially when considering an abstract domain like mathematics.

The term intuition also has special connotations in particular domains. One speaks of "moral intuition" which would represent an *a priori* knowledge of the notions of "right" and "wrong" (Wild, 1938, p. 131).

In the philosophy of Benedetto Croce intuition plays an essential role in aesthetic feelings. According to Croce, beauty is not a property of Nature. It is rather the product of a specific kind of selection and synthesis which is accomplished by the human mind through intuition. As a matter of fact, in Croce's view intuition is always associated with the sense of beauty, because intuition is always associated with unity in a multiplicity of appearances (See Wild, 1938, pp. 39—49).

Since to some philosophers intuition is the way to reach the essence, the absolute truth, a natural consequence would be to consider intuition as the way to approach divinity. Mystical and generally religious intuitions have often been discussed in philosophical and theological works (see Wild, 1938, pp. 97—114). Let me also mention the use of the term intuition as related to professional capacities. A physician, an engineer, a politician, a psychologist etc. may be said to be able to use his or her intuition in solving complex professional problems: the solution seems to appear promptly only on the basis of an apparently summary evaluation.

Let me mention the example of Berne, a psychiatrist who has elaborated a theoretical approach to intuition on the basis of his professional experience. According to Berne a specialist becomes able, as a result of practice, to make correct, global, professional evaluations by resorting to a great variety of cues about which he is, in fact, not aware (*cf.* Westcott, 1968, pp. 42—44).

Related Terms

What complicates the domain of intuition still further is that many other terms are used in reference to the same category of phenomena.

Sometimes people use the term *insight* for indicating a sudden, global rearrangement of data in the cognitive field which would allow a new view, a new interpretation or solution in the given conditions. The terms *revelation* (especially in religious contexts) and *inspiration* (in artistic matters) are also used, sometimes, as synonymous with intuition (at least with some of its meanings).

Very often, "common sense", "naive reasoning", "empirical interpretation" are used in reference to forms of knowledge which may also be considered as equivalent to intuitive knowledge.

M. Reuchlin, a well known French psychologist, has published a very interesting paper devoted to what he has called "la pensée naturelle" (Reuchlin, 1973). "Natural thinking" possesses qualities which distinguish it from formal reasoning, but which play an essential adaptive function: immediacy, concreteness, capacity for sudden and global evaluations.

Piaget uses the French term "self-evidence" in a sense which is very similar to intuitive acceptance. For instance, he writes: "the Michelson and Morley experiment demolished the self-evidence of absolute and universal time" (Beth and Piaget, 1966, p. 194).

Related Areas of Investigation

The domain of intuition and the different and contradictory meaning to which it refers are related to a great variety of cognitive investigations. Let us remember some of them: *Problem solving* (illumination, heuristics, anticipatory schemas etc.); *Images and models* (intuitive representations, intuitive models, intuitive didactical means, thinking in images etc.); *belief* and *levels of confidence*; *developmental stages of intelligence* (Piaget has described intuitive thinking as a preoperational stage).

The Contradictory Domain of Intuition

The attempt to find a common definition for this great variety of meanings, features and connotations seems to represent an impossible task. Intuitive knowledge seems to cover the whole domain of cognition. In Ewing's view, even the formal syllogistic strategies have not, ultimately, any other basis than the intuitive belief in the legitimacy of the restrictive structures.

Wild has described intuition as: "... a plant of confused and intricate growth which has wound its tendrils round many noble trees and mingled its roots with those of the brightest flowers and most ineradicable weeds in the philosopher's garden". (Wild, 1938, Preface). Why then not give up and simply eliminate the term intuition from a scientific vocabulary? It seems that, for each of its virtual meanings, there exist other, more specific, terms like common sense, understanding, belief, guess, insight, and indeed, no psychological textbook, as far as I know, has included intuition among the basic concepts with which it deals. Despite this, the term intuition and its derivatives appear very often, even in psychological descriptions, but without conferring on it a formal, scientific status. Intuition is used rather as a common sense term or as a primitive notion.

In fact we are posing, in this descriptive analysis, threc different but related questions:

Have these apparently very different phenomena termed as intuitions, some basic, common, features or some "family resemblance" which would justify the acceptance of intuition as a definable concept?

If the answer to the above question is affirmative, how is it possible that, despite these common features, the term intuition reveals such contradictory connotations (as, for instance, the highest, the perfect form of knowledge on the one hand; and an unreliable, potentially misleading, form of knowledge, on the other)?

How is it possible that such a confused, hazy term reappears persistently again and again with a pre-eminent role in many important domains like philosophy, science, mathematics, ethics, art, religion?

Let me suggest an answer to the first question. There is a basic common feature which, despite striking differences, allows the various meanings to be related in a common conceptual structure. *Intuitive knowledge is immediate knowledge*; *that is, a form of cognition which seems to present itself to a person as being self-evident.* Therefore, intuitive knowledge may appear, in some texts, as being similar to *sensorial* (perceptual) knowledge. But, at the same time, intuition, as an immediate cognition, may be the source of religious revelations, of artistic inspirations, of scientific illumination etc. In all these instances, one deals with apparently *immediate* forms of cognition.

Why, then, besides this common feature — immediacy — have so many different, contradictory properties been attached to the term intuition (second question)? Why, despite this, has the term intuition survived so long in such a variety of domains and with such a variety of contradictory connotations (the third question)? Would it not be fair to admit that the unique common property — immediacy — is, by itself, too poor for characterizing unequivocally a scientific concept?

THE NEED FOR CERTITUDE

Intuition and Belief

My explanation of the persistent use of the term intuition in many fields and despite the apparent contradictions to which it seems to lead is that intuition expresses, beyond its phenomenological, psychological changing appearance, the natural, almost instinctual belief of every human being in the existence of some ultimate, absolutely reliable, certitudes. In a world of potentially misleading uncertainties our practical decisions cannot rely only upon indirect inferences, on theoretically based suppositions. We feel the fundamental need "to see" with our mind, as we see with our eyes. In order to survive, we have to act in accordance with a given, credible reality. Therefore, *credible* is synonymous with *behaviorally meaningful*. Non-speaking animals are not bothered by credibility. But language and reasoning have produced a breach in this naturally united structure: cognition and behavior. By way of reasoning we know infinitely more than we know through direct perceptual representations. But, as an effect of indirect (conceptual) forms of knowledge, uncertainty becomes a habitual presence in our decision making processes. In order to overcome it a new form of certitude has been invented, corresponding to the symbolic, indirect forms of knowing. This is formal, logically based, certitude. However, the conceptual, logical structure represents a closed system, *a system which controls only its own internal (mental) products.* Its certitudes may have or may not have some practical relevance. What logic offers by itself is not an absolute, practically valuable certitude but *a conventional form of acceptance.*

It is *the need* for a behavioral, practical, non-conventional, implicitly meaningful certitude which creates the almost instinctual belief in the existence of such ultimate certitudes and, consequently, the *quest* for them. It was probably Descartes who best expressed this view: If knowledge is always the product of an active mind, one has to find in the mind itself the criteria through which a certain truth may be distinguished from uncertain appearances.

It is the basic need for unshakeable, self-sufficient certitudes which in our opinion is expressed in the perpetual (very often unconscious) tendency towards direct (directly credible) evidence. It is the *same need* which manifests itself in religious revelations, in the scientists' quest for intuitive models or in the mathematician's endeavor to "see" — at *the end of a tremendous analytical effort* — the solution to a problem as a unique, global, directly acceptable, intrinsically meaningful structure.

Apparently, introspectively — and this view has been consistently expressed by Descartes and Spinoza — one may get the idea that intuitive revelations (experienced as an intrinsic belief) represent the absolute source and guarantee of certitude.

It is the absolute need and the almost instinctual quest for certitude which, historically and psychologically, have shaped this particular type of information processing. Disparate or incomplete data agglutinate themselves through it in apparently coherent, consistent, compact, intrinsically credible structures. They appear subjectively as directly reliable landmarks, indispensable for the continuity and firmness of an efficient mental or practical activity.

The need for certitude and the history of mathematics

As paradoxical as it may appear, it is mainly as an effect of the scientific endeavor towards rigour, in the history of science, that the rich implications of intuitive knowledge have been revealed and described.

It is by striving to render explicit and to purify the formal, the deductive structure of science that scientists and philosophers have discovered the fundamental effects (both positive and negative) of intuitive mechanisms in understanding, solving, inventing and learning.

The contribution of mathematicians has been the most significant, probably because mathematics, by its very nature, is the most suitable for reaching an axiomatized structure. It is in the course of mathematical thinking that the qualities of a formal, ideal model on one hand and the concrete, psychological constraints on the other, appear so sharply contrasting.

While trying to define the concepts used and to build deductive structures, mathematicians have to take maximum care *not* to rely upon intuitive, implicitly accepted, evidence. Consequently, they have to identify the pitfalls represented by intuitively accepted concepts and statements. The problem of evidence has intervened in the history of mathematics in two important circumstances. Trying to build a deductive, logical structure mathematicians had, first of all, to accept a group of initial statements. The criterion used was that of (apparent) self-evidence: if one has to accept some initial, unproved statements as starting points, it is clear that one tries to choose them among such statements which *may* be accepted without proof. This is what Euclid tried to do when choosing his postulates and axioms.

The second circumstance refers to the efforts made by mathematicians to avoid, so far as possible, the misleading effects of (apparently) evident statements. If one follows the history of mathematics one gets the picture of this dramatic struggle of the human mind towards absolute, unconditional truth. The concern of mathematicians to create a rational, self-consistent, system is already reflected explicitly in the works of the Greeks — mathematicians and philosophers. Plato, Aristotle, and Euclid had a clear understanding of the distinction to be made between directly acceptable principles and axioms and those properties which have to be proved. The history of mathematics is, in fact, the history of the endeavours to achieve this programme.

Maintaining intrinsic credibility: an example from the history of physics

As has been discussed above, the historical development of formal structures, with their own form of certitude, does not remove the need for intuitive certitude.

Let us consider for instance, the concept of a *field* in physics. Two bodies, *A* and *B*, exert on each other a force of attraction called gravity. How is this attraction produced? How does the gravitational force propagate itself in space from one point to another? It was assumed, when such influences at a distance were discovered, that some intervening medium must exist which would represent the material support for this propagation. The idea that a body *A* may produce effects on body *B* — and *vice versa* — without any relating medium existing between them, without "something" travelling from one to the other — that idea is *in itself* unacceptable. In a letter addressed by Newton to the Rev. Richard Bentley he clearly expressed this view:

"That gravity should be innate, inherent, and essential to matter so that one body may act upon another at a distance through a vacuum and without the mediation of anything else . . . is to us so great an absurdity that I believe that no man who has in philosophical matters a competent faculty of thinking can ever fall into it". (*cf.* Parsegian, 1968, p. 377).

As a matter of fact, the early interpretations of the field concept assumed the presence of some medium able to convey the influence from one body to another.

Even *after* being taught the accepted view that no intervening medium is necessary, one continues to think about the gravitational interaction as if "something" exists which mediates this transport. One may not be aware of the existence of such an implicit representation but it continues to act tacitly and to influence the ways of reasoning.

The best argument I may produce is the creation of the concept of *field* itself. One accepts nowadays, theoretically, that the force acting between two separate bodies could act *at a distance* without any intervening medium. But such an interpretation *is not credible in itself.* It lacks the convincing evidence of a practical representation. It is the concept of field which replaces the idea of an intervening medium.

Although it seems to eliminate the notion of a travelling or transmitting substratum, the field concept manages to smuggle discreetly a behaviorally credible representation into the domain of theoretical physics.

The notion of field certainly gets a theoretical, formal status in physics which, strictly speaking, does not contain any explicit reference to a substantial intervening medium. But psychologically, it has the quality of being intuitively manageable. One may "see" the field — the distance is no longer totally empty. Maxwell has put into it "lines of force" emanating from charges of one polarity and terminating on charges of an opposite polarity. Thinking in pictures is not intuition itself, but it may increase the direct, intrinsic credibility of a concept and this is the basic, natural tendency in every information processing activity even when the corresponding formal structure may appear to be conceptually irreproachable.

The Crisis of Certitude

The scientific community began to realize that self-evidence is not an absolute guarantee of truth, that what today appears as being an absolute property of a category of phenomena may be rejected tomorrow as an incomplete or even incorrect description of the relevant reality. Truth itself has acquired a more and more relativistic connotation.

The Copernican revolution, the non-Euclidean geometries, the special and the general theories of relativity, the findings related to the Cantorian concept of actual infinity, etc. — all these ideas and representations have contributed to the notion that self-evidence (i.e. intuitive evidence) is not synonymous with certainty. More and more non-intuitive or counter-intuitive concepts have invaded science and mathematics. A continuous function without a derivative has no intuitive meaning. The statement that the set of even numbers is equivalent to the set of natural numbers which is equivalent to the set of rational numbers, has not got an intuitive meaning. What is the intuitive meaning of $a^0 = 1$ or of a division like $(2/3) \div (7/12)$?

It is evident that one cannot spend 5 dollars when one has only 3, but mathematically one may write $3 - 5$ and ask for a reasonable solution.

For a very long time the concept of actual infinity has been rejected because it leads to the logically unacceptable statement that a set may be equivalent to one of its proper subsets. Since Cantor one accepts the concept of actual infinity and one rejects (or at least has to clarify) the apparently self-evident statement that the whole must always be bigger than each of its parts.

The scientific community started to consider direct evidence — expressed by various intuitively accepted statements and representations — as being potentially misleading. Instead, it was the logical form of certitude which became the ultimate stronghold of scientific conviction.

From its royal position as the absolutely credible form of knowledge (Descartes, Spinoza) intuition has become, for many philosophers, scientists and mathematicians the primitive, the less than credible, the very probably misleading source of knowledge.

If intuition had been only one kind of knowledge among others, its role in a scientific endeavor would have been diminished and finally it would have disappeared from scientific debates, as a problem settled once and for all. But, as we know, this has not happened and intuition reappears time and again, praised or blamed, in philosophical studies, in works devoted to the foundations of mathematics, and in educational theories (See for instance the interesting book of Seymour Papert, 1980).

The simple explanation of this apparently contradictory situation is that intuition — as we have already mentioned — does not represent only a category of knowledge among others, accepted or banned. It expresses a necessary attitude deeply rooted in our adaptive behavior. Mathematicians and scientists continue to discover that concepts which have previously been taken for granted as self-evident have sometimes to be abandoned. Nevertheless they continue to use, consciously or not, intuitive, potentially misleading models. And the debates about the foundations of mathematics continue to refer to the role of the intuitive approach *versus* the formal, strictly deductive one.

The work of Brouwer is extremely significant in this regard. After the emergence of the non-Euclidean geometries, after the essential changes produced by the Cantorian approach, after the fundamental works devoted to the axiomatic method, and despite all these, Brouwer claimed that a true mathematical approach must be intuitionistic!

We shall return to this point later on, but let me cite Hermann Weyl in this respect: "Brouwer", says Weil, "opened our eyes and made us see how far classical mathematics nourished by a belief in the absolute that transcends all human possibilities of realisation goes beyond such statements that can claim real meaning and truth founded on evidence" (Kline, 1980, p. 235).

And Weyl is no exception. Morris Kline cites a long list of famous mathematicians who, in the last century, emphasized the role of intuitive acceptance in mathematical reasoning (Kline, 1980, pp. 306—327).

Nobody today would claim seriously that it is time to return to Aristotelian physics, to the concept of phlogiston or to the practical geometry of the old Egyptians.

But nobody seems to be really surprised that many mathematicians continue to claim overtly that the final assessment of the validity of a mathematical statement — or even of a proof — is based on the feeling of subjective evidence — exactly as Descartes did three hundred years ago. You may be perfectly aware of what has happened in the history of science and of mathematics to the notion of self-evidence, that many errors have been made on the grounds of absolutized self-evidence, and yet, looking for an ultimate rampart for the defense of your certitudes, in the inextricable mixture of logical arguments you go back to your personal, very personal, feeling of evidence. This kind of intellectual, I would say very honest, duplicity, seems to me to be at the same time absolutely surprising and absolutely natural.

Intuition is not the primary source of true, certain, cognition but *it appears* to be so because this is exactly its role: *to create the appearance of certitude, to attach to various interpretations or representations the attribute of intrinsic, unquestionable certitude.* As I said before, without a minimum of such apparently absolutely safe stones under our feet, no behavior — practical or intellectual — would be possible.

While writing these lines I am aware, in principle, that many of my assertions have been taken uncritically for granted by myself. And yet I go on believing in every sentence I am expressing. The whole process would have stopped very soon if I had committed myself not to take for granted, without complete control, any of the assertions or inferences made here, including those which are apparently self-evident. Descartes only made explicit what is implicitly accepted in every reasoning activity: one tends irresistably to accept some of the arguments as essentially certain, and the criterion tacitly used is that of self-evidence (or, in other terms, clear and distinct).

Because of the imperative need for implicit certitude as an absolute component of a normal, practical, or mental activity, and because self-evidence is behaviorally the ultimate criterion for certitude, we continue and we shall always continue to fabricate apparently self-evident representations and interpretations. And this is the function of intuition. It does not follow that intuition is necessarily a bad advisor. Intuition summarizes experience, offers a compact, global representation of a group of data, helps overcome the insufficiency of information, introduces behaviorally meaningful interpretations in a reasoning process, and thus confers on the mental activity, the qualities of flexible continuity, of firmness and efficiency which characterize an active, adaptive behavior. But, at the same time, intuition remains a potential source of error because it does not represent a simple duplicate of practically given conditions. Its role is to offer behaviorally meaningful representations, internally structured, of intrinsic credibility, *even if these qualities do not, in fact, exist in the given situation.* It is highly possible that the process of rendering intuitive will produce a distorted representation of the original reality and the predictions made could be totally or partially wrong.

The extremely complex picture one gets when considering the concept of intuition, with its various contradictory attributes, meanings and connotations would appear much more consistent if one accepts the above interpretation.

In a first phase, in the history of science and philosophy the main tendency was to consider intuition as being a source or a method for obtaining absolutely trustworthy knowledge. But more and more findings were accumulated which pointed to the fact that theories and representations, previously considered as being eternal and absolute, may have to be abandoned and that different, counterintuitive interpretations should be accepted. Consequently it was argued that intuition has to be blamed for all these blindly-accepted misinterpretations. What has been considered previously as a source of true knowledge became the supposed cause for many errors and distortions in mathematics and science. The historical evolution of scientific thought may explain this sharp shift in attitude with respect to intuition.

But what the historical evolution cannot explain by itself are the complexity and the contradictory aspects of the picture one gets with regard to the various existing interpretations of intuition. *They coexist even after the discovery that self-evidence is an historical category.*

The same author, who praises the axiomatic approach, himself persistently uses intuitive prompts, procedures and explanations in his reasoning endeavor!

Things become clear, if one admits that, while intuition is not the perfectly reliable source for absolute knowledge, it is, nevertheless, the expression of our fundamental need for absolute, intrinsically reliable landmarks in a reasoning endeavor.

One may strive to eliminate every intuitive impact on the course of a mental activity; one may theoretically claim that intuition is a potential danger for any formal, consistent approach; but what one cannot do is to eradicate the organic need for (apparently) absolutely safe landmarks in the course of a reasoning activity.

The variety of meanings and attributes related to intuition and the variety of procedures commonly called "intuitive", represent the diversity of situations in which intuitive attitudes may intervene and the diversity of means through which the appearance of intrinsic certitude may be created.

A PRELIMINARY DEFINITION

In the present work the term "intuition" will be used generally as an equivalent to *intuitive knowledge*; in other terms not as a source, not as a method, but, rather, as a type of cognition. One admits *intuitively* that the shortest way between two points is the straight line, that every number has a successor, that the whole is bigger than each of its parts, that a body must fall if not supported etc.

All these statements are accepted as being immediate and self-evident without feeling the need for a proof either formal or empirical. Self-evidence is, then, a general characteristic of intuitive knowledge. By contrast, the statement: "The sum of the angles of a triangle is equal to two right angles" is not self-evident, is not accepted intuitively.

It is important to emphasize that we distinguish between *perception* and *intuition*. Perception is also an immediate cognition. I perceive the table in front of me. I have no doubt about its existence. I do not need to prove it. But this I would not call intuitive knowledge.

An intuition, in this view, always exceeds the given facts. An intuition is a theory, it implies an extrapolation beyond the directly accessible information.

If one contemplates two intersecting lines one *sees* that the pairs of opposite angles are equal. This is not a theory, it does not require any intuition. But the statement: "Two intersecting lines determine pairs of opposite equal angles" expresses an intuitive generalization. It is the universality of the property which is accepted intuitively.

One may, then, affirm that intuitions refer to self-evident statements which exceed the observable facts. Let me also add the coercive character of intuitive representations. Being (apparently) self-evident, such representations appear, generally, as absolute, unchangeable ones. It is virtually impossible to accept intuitively, for instance, that a set may be equivalent to one of its proper subsets.

Self-evidence also means *globality*. A certain statement accepted as self-evident is also accepted globally as a structured, meaningful, unitary representation. Consider for example the question: "Two liters of juice costs 3 dollars. What is the price for 4 liters?" The intuitive, correct, answer is 6 dollars. But now consider the question: "One liter of juice costs 2 dollars. What is the price to be paid for 0.75 liters?" In this case, the multiplication does not appear as an intuitive (direct, global) solution. As these examples show, the property of globality serves to distinguish *intuitive* and *analytical* thinking.

In this chapter the question has been posed: how is it possible that the concept of intuition has so many, so different, even contradictory connotations? Starting from the concept itself, no plausible explanation may be found. So many different attributes would never have organized themselves spontaneously within one concept. If this had happened as a result of mere ignorance, the concept would not have survived after its inconsistencies had been discovered.

Things become much more clear if one admits that the concept of intuition, though apparently vague and inconsistent, expresses a fundamental, very consistent tendency of the human mind: the quest for certitude. In evaluating chances, in predicting outcomes, in making decisions, one naturally tends to produce representations (either conceptual or pictorial) which offer a high level of direct credibility.

As a preliminary definition, therefore, it is stated that intuitive cognition is characterized by *self-evidence, extrapolativeness, coerciveness,* and *globality.* These characteristics, and others, are elaborated in succeeding chapters, drawing on examples from mathematics and physics.

CHAPTER 2

INTUITION AND MATHEMATICAL REASONING

THE REAL WORLD AND THE MATHEMATICAL WORLD

Let us recall the basic ideas expressed in the first chapter. Cognitions are essentially structural components of any adaptive behavior. This assertion refers to both the representational and the creative aspects of cognition.

In order to meet the behavioral requirements, the information acquired must be converted into apparently well-structured, self-consistent, action-oriented representations of reality. This is the world of objects and events surrounding us. But human beings have invented ways of obtaining information which is not directly available. These include language, logic and reasoning, and also tools and instruments (indeed, many psychologists emphasize the tool-like characteristics of language). By these means, the structural unity between cognition and adaptive reactions is destroyed.

Knowledge, through reasoning, becomes a relatively autonomous kind of activity, not directly subordinated to the adaptive constraints of the behavior of human beings. In the case of mathematics, that autonomy is quasi-absolute. Mathematics deals with ideal objects and ideal operations, ideal means of verification, the meaning of which is totally determined by formally established definitions and rules. The usual adaptive qualities of objective representations: immediacy, self-evidence, self-consistency, direct credibility, intrinsic necessity (as they appear in sensorial perceptions) are absent from mathematical entities.

Instead of the intrinsic credibility offered by real objects and by practically performed operations, mathematics deals with a formally based type of certitude. Instead of concrete objects, mathematics postulates, formally, the existence of abstract entities. Instead of empirical verification, mathematics uses deductive checks through formal proofs. Proven evidence replaces direct evidence. Instead of the intrinsic, given, coherence of empirical realities, mathematics strives to create sets of sentences, the coherence and consistency of which are formally established.

This is a new world, fundamentally different from that of real objects and real events — the world of mental constructs, internally ruled by laws formally stated, the world of mathematics. It is intended to function in an absolute autarchical way: it produces its own objects; it relates them one to the other according to its own principles; it has its specific type of necessity — logical necessity instead of empirical causality; it has its own type of certitude, a kind of certitude which is reducible to formal rigor (and which may not have any practical relevance).

This world of constructs — the world of mathematics — seems to mirror all the features which enable the known real world to function. To be sure, it mirrors them on its own terms, but all the ingredients seem to exist for conferring credibility, consistency, coherence, on this world of mentally produced abstractions.

In other words, the human mind seems to have learned from the basic general properties of empirical reality how to build an imaginary, structured world, similarly governed by rules and similarly capable of consistency and credibility. The fundamental difference is that in the empirical world the constraints (invariant properties and relationships) are implicitly given, while in the formal world every property and every relationship is stated and justified explicitly. The history of mathematics is the history of the human endeavor for shaping a new type of certitude dealing with explicitly postulated entities governed by explicitly, formally-stated rules.

As has already been said, the ideal aim of this endeavor has been, and still is, the creation of a world of concepts which may function coherently in an absolutely autarchic way. This was the dream of the modern, axiomatic approach.

We may now affirm that the complete independence of mathematics, as a closed world of formally postulated entities, has proved itself to be an impossibility. This impossibility presents two aspects. One is the formal impossibility. It became clear — especially as a consequence of Gödel's incompleteness theorem — that formally, logically, a mathematical system can never be absolutely closed; that is to say, it can never possess in itself all the necessary formal prerequisites for deciding about the validity of all its theorems. The contribution of Gödel to this result is very well known. In Wilder's terms: "These results may be roughly characterized as a demonstration that in any number system broad enough to contain all the formulas of a formalized elementary number theory, there exist theorems (formulas) that can be neither proved nor disproved within the system" (Wilder, 1965, p. 270). This conclusion is in fact true for every mathematical system. It is clear today that it is impossible ". . . to develop mathematics in a complete consistent formal system . . ." (Wilder, *ibid.*, p. 274).

The second aspect refers to psychological impossibility. There is today much evidence — both experimental and descriptive — that no productive mathematical reasoning is possible by resorting only to formal means. One may possess all the formal knowledge relevant to a mathematical topic (definitions, axioms, theorems, proofs, etc.) and yet the system does not work by itself in a productive manner (for *solving* problems, *producing* theorems and proofs etc.). This is what mathematicians affirm in their autobiographical and introspective notes and this is what results from cognitive and developmental, experimental studies.

The formal way of knowing — as we have mentioned above — has tried to mirror all the basic aspects of a behavioral adaptive endeavor: the search for certitude, for consistency, for coherency, for efficiency, the search for invariant relationships (which would guarantee the predictive capacity of the information obtained). *Nevertheless, it seems that the whole system remains sterile if it does not keep an intimate contact with its original, authentic, practical sources.*

It seems that the formally based qualities of certitude, coherence, consistency, necessity, etc., *do not possess the same kind of stimulating, convincing and productive capacity as the intrinsic credibility, the intrinsic structurality and richness of real phenomena.* This is what Hilbert himself, one of the great founders of axiomatics, has clearly stated: "Who does not always use, along with the double inequality $a > b > c$, the picture of three points following one another on a straight line as the geometrical picture of the idea "between"? Who does not make use of drawings of segments and rectangles enclosed in one another when it is required to prove, with perfect rigour, a difficult theorem on the continuity of functions or the existence of points of condensation? Who could dispense with the figure of the triangle, the circle with its center or with the cross of the three perpendicular axes? Or would give up the representation of the vector field or the picture of a family of curves or surfaces with its envelope which plays so important a part in differential geometry, in the theory of differential equations, in the foundations of the calculus of variation and in other purely mathematical sciences?" (Reid, 1970, p. 79).

Let us consider, for instance, the statement: "Between two consecutive roots of the derivative of a function there is no more than one root of the function". This sentence is related to the Rolle theorem and is used for solving polynomial equations.

If a pupil is presented with the above sentence together with its proof, he will probably be able to memorize them. He may also be convinced, relying on the proof, that the sentence is correct. But he will not be able to use it and other related sentences as stimulating devices for his mathematical thinking; at best he will be able to use the theorem for solving mechanical, standard problems. The theorems of Fermat, Rolle and Lagrange are interrelated. By merely learning the verbal and symbolic expressions of these theorems and proofs a pupil will not be able to gain a genuinely coherent understanding of the whole set of sentences.

But if one adds an image, as one usually does, to the verbal presentation, things may change radically. One *sees* that between two consecutive extrema one may not have any root or one may have a single root and no more. The existence of two roots, i.e. two points in which the curve intersects the axis, would imply another extremum.

The above example is less trivial than it seems to be at a first glance. Actually, it is not enough to perceive the image in order to understand, intuitively, the theorem: *the psychological problem is to become convinced that always, necessarily, a point travelling from one extremum to a consecutive*

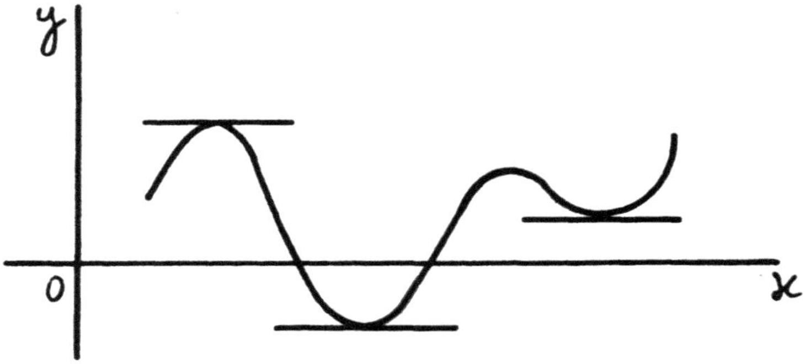

Fig. 1.

one cannot cut the x-axis more than once. And this is much more than "seeing" the graph. The given picture has to appear as one of a multitude of possible graphs of the same category (i.e. only as a particular representative of a class of curves obeying certain conditions); and then this class itself has to appear as a sub-set of a more general category of curves which may, or may not, observe the given conditions (that is, for instance, curves cutting more than once the x-axis). The graph must be grasped not as a final, fixed state but rather *in construction,* that is as a point moving in the $x0y$ plane, reaching a maximum, turning towards the axis, cutting the axis, reaching a minimum, turning again to the axis and cutting it again. Usually, we are not aware of all these "happenings", but without them in the "arrière-plan" of our thinking, the image of the graph will not engender the expected "intuitive" understanding of the theorem, as expressing a *general, necessary relation.* Newton did not consider mathematical magnitudes as consisting of entities which may be as small as one wants, but rather as the result of a continuous movement. According to him lines are not produced by a juxtaposition of parts but by a continuous movement of points.

We are commonly so hypnotized by the striking characteristics of the image itself that we are not aware of the whole process which lies behind the elaboration of an intuitive understanding (in this case, our behavioral identification with the construction of the curve and the events which take place as an effect of this process).

Actually, it took a very long time in the history of mathematics before mathematicians became able to grasp the meaning of a changing rate in the dynamics of a function and to translate it algebraically. Only in the fourteenth century did Nicole Oresme notice that the rate of increase or decrease of a magnitude is slowest in the neighborhood of a maximum or a minimum (Hadamard, 1949, p. 144).

Only three hundred years later, with Fermat, did this observation receive a mathematical meaning and a relation was established between the zero value of the rate of change and the points of maximum and minimum.

To summarize: as has already repeatedly been stated, the essential role of intuition is to confer on the conceptual components of an intellectual endeavor the same properties which guarantee the productivity and the adaptive efficiency of a practical behavior.

Hence, one tends to assign to concepts and formal operations interpretations which are able to fulfill some specific behavioral requirements:

Firstly, such an interpretation has to be *globally, directly representable*, that is to say it has to be able to guide the course of the whole process.

Secondly, this interpretation, although somehow apparently static, has to suggest *a constructive activity* in order to confer a behavioral meaning on the respective representation.

Thirdly, the interpretation *has to exceed the particular given representation*. One grasps — through a unique, limited, incomplete representation — the general principle of construction and thus one is led to expect the continuation or even the achievement of the whole process as something *as naturally, evidently determined*; *as if it were actually given*.

A graph representing a function may become an intuitive tool only if one grasps it as representing not a final, already accomplished state, but rather as an "instantané" of an event actually in progress. (Computer software such as the EUREKA program produced by the ITMA team at Nottingham has enormous potential for extending graphical representation from the static to the dynamic.)

Our point of view is that *one tends always, almost automatically, to produce complementary interpretations of the conceptual structures which, by their very nature, will be able to confer on the concepts used the direct credibility, consistency and intrinsic necessity required by a normal, productive behavior.*

One may not be aware of the existence and of the impact of these implicit interpretations and representations and this is what makes controlling them so difficult.

THE BEHAVIORAL MEANING OF MATHEMATICAL CONCEPTS

Graphs representing functions are in fact an extreme case; they are ideally good models for abstract mathematical relationships. They are an important instrument, very often well controlled by the conceptual structure.

But the common case is that of interpretations which are produced spontaneously, and used tacitly without benefitting from the systematic control of the formal conceptual instance.

A nine-year-old child being asked to compare a point determined by the intersection of two lines with the point of the intersection of four lines, answered that the second point was bigger. There is not any essential difference between this child and the mathematicians who, before Weierstrass, were convinced that every continuous, real function is differentiable at some point at least. In both cases, tacit, spontaneous, uncontrolled interpretations distorted the reactions. If only pure, formally stated concepts had been used, the answers would have been different. Certainly, one strives to identify and reject such apparently ready-made, pseudo-evident representations and interpretations. But, in fact, this aspiration is never, and cannot be, fully successful. Our symbolic behavior, like our practical behavior, needs to rely upon apparently existing, objective, unquestionable "facts".

The mechanisms of reasoning are, to a great extent, beyond direct, conscious control. One may, relatively, control the sources of knowledge, the meaning, the pertinence of various concluding steps of a reasoning activity, but the heuristic activity itself is mostly beyond any systematic conscious control. At this deep-structure level, the mind produces a large variety of means through which the plausible appears as certain, the non-visible as visible, the infinite as comprehensible, the abstract relations as behaviorally interpretable.

Let me analyse another example, the notion of a straight line. It is, obviously, an abstraction. There are no "straight lines" in reality. For a physicist, a straight line corresponds to a light beam. To a pupil, a straight line is a line drawn on a sheet of paper. To a traveller, a straight line means going straight ahead. These representations are physically different. Nevertheless, when using the term "straight line" one may be convinced, before any specific, theoretical, analysis, that one knows perfectly well what one is talking about. The notion of a "straight line" seems to be as self-evident and self-consistent as the desk on which one is writing or the sunshine during a bright day. The notion of a straight line seems to evoke an absolutely clear and obvious meaning. One is naturally convinced that one may go on extending the line indefinitely, that, by following the straight line, one actually uses the shortest path to reach a certain objective etc. All these seem to be absolute, unquestionable properties of the "fact", of the "object" called a straight line, like the fact that the walls in my room are white, or that a certain book on my desk is particularly thick and heavy.

Self-evidence and credibility are naturally implicated in the primitive notion of a straight line. As a matter of fact, there is no such concrete, unequivocally definable reality which would infuse its own self-evidence and objective consistency into the corresponding notion. No such object exists. As an effect of psychological behavioral constraints, we tend to confer on this notion and the corresponding statements — which are far from being clear, self-consistent and self-evident — the qualities of unequivocal evidence and credibility.

Notions and statements may appear to be conceptually evident and intrinsically credible only because they are actually, in certain circumstances, *behaviorally meaningful.*

The straight line is a concept, not an object. It is, I would say, a normative concept. In a definite, practical context, it has a practical meaning. I have no doubt that I know how to draw a straight line or to recognize a straight line, etc. But, as a matter of fact, the concept of a straight line has no absolute objective meaning. Beyond practically defined circumstances, it is a *convention* defined in the frame of a certain group of axioms, *which may be changed.* But via extrapolation (from a behavioral meaning contextually dependent on an absolute universal concept), *one tends to believe in the absoluteness of the concept, one tends to confer on this notion, based on conventions, the absoluteness of a given, objectively existing fact.* In our terminology, this means conferring *intuitivity* on a concept, or, in other words, the concept of a straight line gets an intuitive meaning for the individual.

It is easy to find examples of the same type in the domain of geometry. The notion of a point gets an intuitive meaning by attaching to it the image of a small spot, "as small as we want", although we know perfectly well that a point is a pure concept, that there are no such objects in reality which may be called "points" in the mathematical sense. We mentally manipulate points, lines, geometrical figures as we manipulate objects, though we know very well that all of these are *not* objects. But they have, for us, subjectively, *an intuitive meaning.* The respective concepts and operations produce, in fact, for the individual's intellectual apparatus, the internal consistency, the empirical reliability, the practical "manipulability" which characterize real, concrete objects.

The main attribute of intuitive knowledge, as we have frequently repeated, is the feeling of *direct certitude* and this is produced, first of all, by the impression of self-evidence. A normal mental behavior is possible only if it may rely, automatically, on a number of intrinsically acceptable data — like every normal practical behavior. And this is not possible without believing in the absoluteness of empirical reality. We are not referring to an explicit, philosophically-stated conviction. We refer to an automatic mental attitude, to the fact that the organization of our behavior is entirely based on this spontaneous belief in the absolute existence of the external reality.

Our theory is that mental behavior (reasoning, solving, understanding, predicting, interpreting) including mathematical activity, is subjected to the same fundamental constraints. *The mental "objects" (concepts, operations, statements) must get a kind of intrinsic consistency and direct evidence similar to those of real, external, material objects and events, if the reasoning process is to be a genuinely productive activity.*

An intuition is, then, an *idea* which possesses the two fundamental properties of a concrete, objectively-given reality; *immediacy* — that is to say intrinsic *evidence* — and *certitude* (not formal conventional certitude, but practically meaningful, immanent certitude).

Intuitive representations will not disappear from mathematical endeavors merely because one decides that such representations do harm to the rigor of a formal reasoning process. They will remain because they are an integral part of any intellectually productive activity.

This affirmation is based on the postulate that thinking is a form of behavior. It tends automatically, by its very nature, to preserve those properties and mechanisms which guarantee the productivity, the efficiency, the continuity of any adaptative behavior. Formally based credibility (through a formal proof) may be sufficient for producing a conventional conviction, but it is not sufficient for guaranteeing the genuine progress of a reasoning endeavor.

AXIOMATIC STRUCTURES AND INTUITIVE PROCEDURES

What has been said about certitude may also be claimed with regard to other fundamental properties of a mathematical system. In order to check the consistency of the system, the completeness of the axiomatic basis and the independence of the axioms, one resorts, generally, to a model, to a particular interpretation of the system, although, in principle, such checks might be performed at a pure, formal level. The model offers to the active reasoning endeavor, the *sine qua non* analog of a concrete, particular, objectively given reality.

Theoretically, in order to build an axiomatically structured system, one has to start from a group of axioms and undefined terms. By combining them, one gets the various theorems of the system. In order to prove that the system is consistent, one has to show that no contradictions appear between the various theorems which may be produced, starting from the given axioms. But such a procedure is not efficient in practice; one never knows whether *all* the possible theorems have been produced. Moreover, it is psychologically doubtful that working with, let us say, hundreds of theorems one may be sure that no contradiction among theorems has escaped unobserved.

The practical procedure is to resort to *a model*, a particular interpretation which would appear *psychologically* as a structured reality and with regard to which, contradictions would become salient. When we say *structured* we mean *intrinsically coherent*, as real objects are. It is not possible to accept, simultaneously, that a line l_1 intersects a line l_2 and that the two lines have not a common point. But if one does not resort to any such internally structured interpretation and one deals only with symbols, the meanings of which are defined by axioms, it is possible to overlook the contradiction. In other words, the really feasible way to check the consistency of an axiomatic system is to relate the system to a *model* which would have the fundamental feature of being *intrinsically structured*, similarly to concrete objects.

One may repeat the same practical principle with regard to the method of checking the independence of the axioms of a mathematical system. The axioms of a system are independent if none of them may be deduced from the others.

In principle, in order to prove the independence of a certain axiom in a system, one should produce all the consequences of the other axioms of the system in order to check whether the tested axiom may be found among them. This, again, is not a practical procedure. Instead, one resorts to a model which organizes the ideas and which suggests various possible, deductive, meaningful connections between the chosen axiom and the remaining ones. In Paul Cohen's words: "The most natural way to give an independence proof is to establish a model with the required properties. This is not the only way to proceed since one can attempt to deal directly and analyze the structure of the proofs. However such an approach to set theoretic questions is unnatural since all our intuitions come from our belief in the natural, almost physical model of the mathematical universe." (Cohen, 1966, p. 107).

As a matter of fact, the axiomatic structure is the final state attained by a body of mathematical knowledge *after* the body of knowledge has already been obtained by other means than mere deduction. These procedures refer to heuristic and inductive processes similar to those which intervene in empirical sciences. An axiomatic system may correspond to various particular interpretations or models because of its absolute formality. But, in fact, when building an axiomatic system, one starts from a certain, particular, relatively familiar, internally structured group of concepts (for instance, points and lines as they appear in elementary geometry) and one tries to generate formal statements about this body of concepts. Only afterwards does one check whether the deductive structure obtained is also applicable to other models (see Wilder, 1952, pp. 8—21).

The main idea is that *the same type of mental attitudes and endeavors which characterize an empirical attempt at solution intervene also at the formal level.* And this, despite the fact that the formal thinking contains its own mechanisms for connecting and structuring information, for identifying invariants, for elaborating decisions and producing credible conclusions. But these mechanisms by themselves are no more than a check list, a program for checking the formal validity of the system; it shows what properties to look for, and defines them (consistency, independence of axioms, completeness of the system). But the formal structure *does not* contain, in itself, programs for *selecting* the axioms, for *seeking* for meaningful statements, for *practically performing* the checks for consistency, independence and completeness. All these operations are accomplished in the same way as in any other mental, productive behavior. Therefore, even when dealing with axiomatical structures, the mathematical activity resorts to the intuitive forms of acceptance and extrapolation which may assure its required behavioral firmness, its productivity, its dynamic, flexible consistency! As Raymond Wilder has clearly stated: "Thus, in practice, the concept comes first, the axiom later. Theoretically, this is not necessary, of course. Thus we may say, 'Let us take as undefined terms *aba* and *daba* and set down some axioms in these and univeral logical terms'. With no concept in mind, it is difficult to think of anything to say! That is, unless *we first give some meanings* to "*aba*" and "*daba*" — that is, unless we introduce some concept to talk about — it is difficult to find anything to say at all. And if we finally do make some statements without first fitting a suitable concept to "*aba*" and "*daba*" we should very likely make statements which contradict one another! As we shall see below, the underlying concept is not only a source of the axioms but it also guides us to consistency . . ." (Wilder, 1952, p. 19).

INTUITIVE CONCEPTS AND THE SCIENTIFIC COMMUNITY

One may distinguish different situations concerning the status of intuitive acceptance of various "truths" as they are defined by the scientific community at a certain epoch.

There are situations in which an intuitive acceptance coincides with what is usually accepted as true by the scientific community. At least from a certain age (what that age is, is the subject of controversy among developmental psychologists), people accept intuitively that if $A = B$ and $B = C$ then $A = C$. One accepts intuitively that every number has a successor. It seems evident that from a point outside a line one may draw one and only one perpendicular to that line etc., etc.

On the other hand, one is also ready to accept, intuitively, assertions which, as a matter of fact, are rejected by the scientific community, and *vice versa*. It is difficult to accept that the set of natural numbers and the set of positive even numbers are equivalent. It is impossible to grasp intuitively the meaning of symbols like a^0, a^{-3}, $a^{0.5}$, or the meaning of an operation like $3/7 \div 0.21$. It seems intuitively evident that space possesses absolute, privileged directions: "up" and "down"; "vertical" and "horizontal". It is almost impossible to accept, intuitively, that simultaneity has no absolute meaning etc.

Things appear to be still more complicated if one remembers that terms like "fact", "truth", "evidence" have no absolute value.

A sentence may be accepted as true by the scientific community in a certain period, it may also appear as intuitively acceptable to the layman, and yet one may discover later, in the history of science, that the sentence is contradicted by facts, previously not known, or rejected on logical grounds, previously not taken into account. But the so called "facts" and the "logical reasons" are themselves subject to interpretations which are historically alterable!

Certainly, one has to distinguish between the philosophical problems referring to the historical, relativistic character of "truth", and the psychological problems concerning the relationships between the individual's intuitive acceptance of a sentence and the truth value assigned by the scientific community to that sentence. But the relationships between the three factors considered may give rise to extremely complex and psychologically interesting situations (the three factors being the individual's conception of a phenomenon, the "classical" conception of the scientific community with regard to the same phenomenon and a conception which may appear, at the same epoch, as being the "new", the controversial, the "revolutionary" one). An individual may become subject to the simultaneous pressure of these three distinct, and even contradictory, conceptions, each of them belonging to a different conceptual system. *Intuitively,* space appears to be non-isotropic (i.e. having privileged directions as it appeared to Aristotle). The absolute, empty, continuous, homogeneous and isotropic space of Newton is difficult to accept intuitively. But one gets used to it. After years of mental "practice" such a space may become intuitively acceptable, although the old representation continues, tacitly, to survive. At the same time, one learns about the relativity of space and time, about the dependence of spatial properties on mass parameters. This is not an hypothetical scenario. This is the reality of every student taking a course in physics.

One knows the enormous difficulties encountered by Cantor in his efforts to get his views concerning the transfinite numbers accepted by mathematicians of his time. His letters, addressed to Dedekind, contain dramatic accounts about his own doubts and his own endeavor to clarify his astonishing discoveries for himself and for the mathematical community. The conflict between the intuitive constraints and the logical implications of his analyses is sometimes expressed in his letters in pathetic tones.

Cantor relates, in one of his letters, that he once asked a group of colleagues whether a unidimensional continuum of an infinite set of points may be placed into one-to-one correspondence with a p-dimensional continuum of an infinite set of points. (For instance whether the points of a segment may be placed into one-to-one correspondence with the points of a square or a cube). And he writes: "Most of those to whom the question was addressed were very surprised that I asked such a question, because it was evident that p independent coordinates are necessary in order to determine the position of a point in a p-dimensional multitude". And Cantor continues: "But by analysing thoroughly the question we must admit that it is at least necessary to prove that the negative answer has to be accepted as "evident". As far as I was concerned I belonged to those who considered the negative answer *as plausible* until very recently when I reached the conclusion, after a series of complex reasonings, that the answer is affirmative without any restriction." (Cavaillès, 1969, p. 168). At that moment, Cantor did not yet possess a proof! Nevertheless he was already convinced that the answer must be *affirmative.*

The impossibility of the equivalence between two sets of points belonging to continuous multitudes with different numbers of dimensions is accepted naturally as a primary truth. Those to whom this problem was addressed were simply surprised that somebody had had the idea of raising the question.

Several aspects should be emphasized. First, a solution to a mathematical problem was accepted "blindly" as certain, by a group of mathematicians (the main concern of whom should have been first to check the validity of the sentence they were using and accepting). Secondly, that in this case they did not refer to abstract duplicates of real objects (from which their confidence could have been derived) but rather to a fantastic world of an infinity of non-dimensional entities (points). *Both concepts — actual infinity and non-dimensional elements — have absolutely no intuitive meaning, no practical correspondent.*

The sentence *rejecting* the equivalence between continuous multitudes with different dimensions is mere speculation, a mere guess, before an explicit proof is found. *Despite all these, the respective speculation was accepted as expressing an absolute truth.* It was accepted not only by some laymen or by some minor professionals but by the scientific community as a whole in a certain period!

This is intuitive knowledge — a kind of knowledge which is not based on sufficient empirical evidence or on rigorous logical arguments and, despite all this, one tends to accept it as certain and evident.

It is worth remembering that Cantor himself, after having obtained a proof supporting the equivalence of the respective sets — *a proof which in his eyes was unquestionable* — was still worried with regard to the validity of the theorem! In a letter addressed to Dedekind on the 29th of June 1877 he wrote the famous sentence: "I see but I do not believe" (Cavaillès, 1969, p. 169).

The fact of obtaining a proof supporting the equivalence should have strengthened his conviction. But four days *after* having written to Dedekind that his conviction is without restriction and after having obtained the formal proof, he still seemed shocked by his discovery. It seems that, while most of the mathematicians of his time were trapped in *one* intuition, Cantor was trapped in *two* contradictory intuitions: the old, "natural" intuition, according to which two continuous sets of points having a different number of dimensions cannot be equivalent and the new, the Cantorian intuition claiming the equivalence of the two sets. The fact of obtaining a proof supporting the second view (the equivalence of the two sets), does not solve the subjective conflict. Cantor is no less anxious and worried *after* obtaining the proof than he was before. He presses Dedekind to help him to overcome the difficulty: ". . . the result I have informed you about, appears to me so unexpected, so new that I would not be able to find my spiritual quietness before receiving, dear friend, your opinion about the exactitude of my finding." (*Cf.* Cavaillès, p. 169).

Of course, this is an extreme example. Very often one deals in mathematics with entities and operations which have some practical correspondent. But this example illustrates the fact, fundamental to our discussion, that the feelings of obviousness, of immanent certitude and consistency are not necessarily based on what may be considered as being full, consistent — empirical or logical — evidence, even in highly educated persons, even in experts of the highest standards. *The drama of certitude is played at another level, with other mechanisms than those related to the dynamics of algorithms and logical connectives.*

CHAPTER 3

INVESTIGATIONS IN OVERCONFIDENCE

OVERCONFIDENCE AND INTUITIVE BIASES

One of the main propositions of the present work is that intuition expresses the fundamental need of human beings to avoid uncertainty. Promptly adjusted, well adapted reactions of a person to given circumstances are possible only if the perception of the respective reality appears to him, automatically, as coinciding with reality itself. Doubts, hesitations are useful only when referring to aims which are not directly involved in the current flow of behavior. When crossing the street, you have to believe absolutely in what you see — the approaching cars, the various distances etc. — otherwise your reactions will be discontinuous and maladjusted. Analogically, during a reasoning process, you have to believe — at least temporarily (but absolutely) — in your representations, interpretations or momentary solutions, otherwise your flow of thoughts would be paralyzed. It is this type of belief that we call an intuition. Cognitive beliefs, elaborated and confirmed repeatedly by practice, may acquire an axiomatic character.

While perceptions of given realities are direct, and hence normally correct, mental representations, ideas, hypothetical solutions may be biased, distorted, incomplete, vague or totally wrong. In order to believe, nevertheless — at least temporarily — in such mental productions, a degree of overconfidence is needed.

That is to say, you very often have to be more confident in your interpretations and cognitive decisions than would be warranted by an objective evaluation of them, in order to keep the reasoning process going. *After* a certain (provisional) solution is reached, one usually initiates some sort of analysis and verification process. But as long as representations and interpretations constitute active links in the flow of mental behavior doubts must not interfere. Doubts may arise about the conclusion obtained as an effect of a retroactive, analytical control. We then tend to project, retroactively, a doubtful character on the various chains of the reasoning process. But, in fact, during the reasoning activity, we behave as if at every moment we believe in our ideas without hesitation. As a matter of fact even after a certain decision is taken one frequently tends to be overconfident about the conclusion reached and to overlook possible counter-arguments. The need for verification usually is less honored than it should be.

As I said, intuitions are specifically those cognitions in which overconfidence plays an essential role. This does not mean that intuitions are always wrong. It only means that we are inclined to admit, with a feeling of absoluteness, statements which are objectively only weakly supported by empirical data or logical arguments.

An experimental argument in favor of the above hypothesis (although an indirect one) is the finding, frequently confirmed, that people generally tend to overestimate the accuracy of their own knowledge and interpretations. This finding is not necessarily related to research on intuition. Very often it deals with factual data.

What this type of finding proves is the existence of a basic tendency of human beings towards overconfidence with regard to the accuracy of their own cognitive decisions. Let me quote some experimental results along these lines.

A group of psychologists was presented with a case of maladjustment of a person who had never been psychiatrically hospitalized. On the basis of the information provided, the subjects (the psychologists) had to make evaluations concerning customary behavior, patterns of attitudes, interests and typical reactions of the person. The information was provided over four successive stages referring to the main periods of the person's life (from childhood to college years). The subjects had to make evaluations after each set of information was provided. They also had to evaluate the accuracy of their decisions in terms of expected percentages of correct decisions.

It was found that the accuracy of the decisions did not improve significantly as a result of the increasing amount of information provided over the four stages. As a matter of fact, the average (final) accuracy was less than 28%, while the probability of providing correct answers by mere chance was 20% — an insignificant difference (Oskamp, 1982, p. 291). At the same time the confidence of the subjects in their own cognitive decisions increased steadily from 33.2% at the first stage to 52.8% at the fourth stage. At each of these four stages the subjects displayed overconfidence with regard to the accuracy of their judgments. It is important to emphasize that the subjects did not get any feedback concerning their cognitive decisions during the experimental session. Nevertheless, they became more and more confident about their reactions.

This finding seems to be of fundamental importance in connection with the mechanisms of intuition. As one continues to accumulate information with respect to a certain topic one tends to organize it in intrinsically credible structures. The fact that the degree of credibility is growing with the quantity of information obtained *independently of any feedback* means that, automatically, a certain selection of the incoming information is made in reference to some primary schema. This schema tends automatically to reinforce itself by selecting and interpreting the sequences of new data coming in, so as to increase the self-consistency and the intrinsic credibility of the original conception. In other words, the main criteria for selecting the new data are determined by their contribution to increasing the consistency and the credibility of the initial intuitive interpretation rather than by the objective novelty and richness of these data.

Let me quote Stuart Oskamp himself: "Regardless of whether the task seemed strange or the case materials atypical the judges' confidence ratings show that *they became convinced of their own increasing understanding of the case*. As they received more information, their confidence soared. Furthermore, their certainty about their own decisions became entirely out of proportion to the actual correctness of those decisions". (Oskamp, 1982, p. 292). Popper relates an extreme case of the same phenomenon: "As for Adler, I was much impressed by a personal experience. Once, in 1919, I reported to him a case which to me did not seem particularly Adlerian, but which he found no difficulty in analysing in terms of his inferiority feelings, although he had not seen the child. Slightly shocked, I asked him how he could be so sure. 'Because of my thousandfold experience', he replied; whereupon I could not help saying: 'And with this new case, I suppose, your experience has become thousand-and-one-fold'." (Popper, 1963, p. 35).

Since the subjects in these examples were psychologists one might argue that the conclusions drawn refer specifically only to that profession. Actually, the tendency towards overconfidence has been described by numerous authors. Most of these studies did not refer to what may be called *intuitive evaluations* but rather to information the subjects were supposed to possess. Nevertheless, the manifest overconfidence of the subjects with respect to the correctness of their own cognitive reactions supports the idea that, basically, generally, people tend to overlook the shortage of information with regard to a certain interpretation in order to increase its immediate, apparent, credibility. One tends automatically to increase the degree of confidence beyond what is justified by the genuine knowledge one possesses. Our standpoint is that when this phenomenon intervenes with respect to an interpretation, a conception or a solution (and not with regard to raw facts or data) one gets what is commonly called an *intuitive cognition*.

It is interesting to analyse in some more detail the mechanisms of overconfidence. Let me mention first some other studies in the same field. Generally, in this class of studies, subjects have to answer a number of questions and they are asked to evaluate, by some clearly defined procedure, the degree of certainty of their answer. For instance, Fischhoff, Slovic and Lichtenstein (1977) used the following procedure. Subjects were presented with a question stem which they were asked to complete — for example: "Absinthe is . . .". After writing down an answer, the subjects estimated the probability that their answer was correct using a number from 0.00 to 1.00.

In one alternative format, subjects were asked to assess the probability (from 0.00 to 1.00) that simple given statements were correct. For instance: "What is the probability that absinthe is a precious stone?"

In a second alternative format, subjects were asked to choose the correct answer from the two that were provided. Again, the subjects had to assign a probability of correctness to the chosen answer (from 0.5 to 1.00).

The third alternative format was similar to the second. The only difference was that the experimenter randomly selected one of the two answers and the subjects had to estimate the probability that the choice was correct.

In a different experiment the same authors asked the subjects to express in odds their confidence in their solutions. It was supposed that in this way the subjects would have more possibility of expressing extreme confidence. An open-ended scale was presented of this form:

10:1	100:1	1000:1	
10000:1	100000:1	1000000:1	etc.

An answer like 1:1 means that one is equally likely to be wrong or right, while 1000:1 means that one is a thousand times more likely to be right than wrong.

In all these experiments, one compares statistically the degree of confidence expressed by the subjects with the percentage of answers actually correct. For instance (experiment 2), it was found that in the cases in which the subject indicated 100:1 odds of confidence (99% confidence) they were in fact correct with regard to only 73% of answers and certainly this points to a degree of overconfidence.

According to Fischhoff *et al.* (1977) extreme overconfidence may have various sources. People are not critical enough about the validity of their own inferences. Pitz considers that people tend to overlook the uncertainty associated with the early stages of inference (Pitz, 1974; *cf.* Fischhoff, 1977). Another aspect is related to the retrieval process. People are not aware, generally, that memory is based on a reconstruction process rather than on a simple duplication. As a result, we tend to confer a degree of authenticity on our memories, which is higher than is really the case. Memories may be distorted by this reconstructive process under the influence of structural, social, motivational factors. "If people are unaware of the reconstructive nature of memory and perception and cannot distinguish between assertions and inferences . . . they will not critically evaluate their infered knowledge" (Fischhoff *et al.*, 1977, p. 539). Accordingly, being unable to evaluate the validity of their answers, people will tend to be overconfident. The basic theory of Fischhoff, Slovic and Lichtenstein is that overconfidence is caused mainly by a natural lack of awareness of people about the mechanisms of memory and judgment.

But, while this theory explains why overconfidence is possible, at least in some circumstances, it does not explain why it actually appears. One would have expected that as an effect of accumulated negative feedback during the historical development of mankind, successive generations should have become less and less overconfident about their own primary cognitive reactions. One would have expected that the degree of overconfidence would be smaller in scientists than in laymen, in adults than in children etc. One would have expected that across the history of science, intuition would have played a decreasing role (both negative and positive) in scientific endeavors, predictions and interpretations. But none of this happened.

It is true that, as has been shown, people sometimes may even become underconfident as the degree of accuracy of their answers increases (above a degree of 80% of accuracy a tendency towards underconfidence appears) (see Lichtenstein and Fischhoff, 1977).

But these findings are related to specific domains of competence and not to the general attitude of people towards their cognitive performances.

Let me give an example. If a mathematician is presented with a theorem he will necessarily ask for a proof even if he intuitively may feel that the theorem is correct. The same mathematician will be ready to display an impressive quantity of educational theories and didactical solutions with the highest degree of confidence, without any empirical control and even without feeling the need for such a control. The lack of psychological and educational knowledge may explain why that mathematician may be overconfident with regard to the educational domain, but it does not explain why he is really so. In fact, he displays overconfidence with regard to many other domains as well, except mathematics itself. May one generalize the case of mathematicians to suggest that, generally, experts are better calibrated with regard to knowledge and predictions in their own domain of expertise? The term "calibration" used here means the degree of agreement between the probability of being true assigned by a person to his own cognitive decisions (or to a certain given statement) and the real proportion of situations in which his or her decisions are confirmed. For instance, let us consider the situation in which a person has assigned the probability of 0.7 of being true to 100 independent items from a questionnaire. If 70% of these propositions are really true one considers that the person is perfectly calibrated. But let us come back to the role of expertise. In a review, Lichtenstein, Fischhoff and Philips (1982) described a number of findings referring to the effect of expertise on the degree of calibration. The data obtained by various researchers are not consistent. It has been found that students taking a practice mid-term examination were 98% correct when assigning a 1.0 probability of being correct to their answers; they were correct in only 0.5% of their reactions when assigning 0 to their answers. In other words, these students were very well calibrated (Sieber, 1974; *cf.* Lichtenstein *et al.*, 1982, p. 321).

On the other hand, Christensen-Szolansky and Bushyhead (1981) reported that a group of physicians were asked about the probability of a pneumonia diagnosis for 1531 patients who were examined because of a cough. Their calibration was very bad. For a level of confidence of 0.88 the proportion of patients actually having pneumonia was less than 0.20 (Lichtenstein *et al.*,

1982, p. 321). With regard to future events the same severe overconfidence has been found (*ibid.*, p. 323).

It has also been found that overconfidence is very high with very difficult tasks. In other words, subjects tend to keep a certain level of confidence even when the task is practically impossible.

Attempts have been made to improve calibration by stimulating motivation or by special training. It has been found (Sieber, 1974) that more motivated students show significantly worse calibration — that is, greater overconfidence — than less motivated students (Lichtenstein *et al.*, 1982, p. 320). Koriat, Lichtenstein and Fischhoff (1980) have found that, by special training, calibration may be improved. Subjects have been asked to solve a number of items by choosing, for each item, a certain answer. One group was asked to write reasons supporting their answers and another group was asked to write reasons which would contradict their answers. It was found that only the group asked to write contradicting reasons showed improved calibration. "This result . . . suggests that an effective partial remedy for overconfidence is to search for reasons why one might be wrong" (Koriat *et al.*, 1980).

The studies in overconfidence — although generally not explicitly related to intuition — are nevertheless highly relevant to that domain.

Firstly, as has already been mentioned, overconfidence seems to represent a very general phenomenon. This fundamental finding seems to support our theory concerning the nature and the role of intuition. We have argued that one tends naturally, almost instinctively, to resort to solutions, conceptions, interpretations which best display a high degree of *intrinsic* (apparent) credibility. The reason for this, it was said, is that firmness and continuity in the very course of an activity may only be possible if that activity is guided by apparently firm, highly credible representations. *Overconfidence seems then, to be the remedy for compensating for the virtual lack of confidence which would have been, necessarily, generated by an accurate evaluation of our ignorance.* We live in a world of incertitude despite the fact that it seems to be governed by eternally valid laws. It has been argued here that intuitive components of cognition are surviving in individuals and in the history of science, despite the steady growth of rationality, because this is the solution to our essential behavioral need for intrinsic certitude.

A second important finding supporting the same conception is that overconfidence may be corrected by revealing to a person the reasons why he is *not* correct with regard to some of his answers, rather than helping him to justify his correct answers.

This means that overconfidence is not an automatic product of ignorance or incompleteness of information. Overconfidence is attained by a selection activity which is aimed to preserve, automatically, those data which seem to support a certain conception and at the same time, to ignore those contradicting it.

A physician who assigns a certain diagnosis to a patient is naturally overconfident in his choice, because in order to be able to show firmness in his professional decisions he must, at least temporarily, ignore or minimize the significance of the possible counter arguments with regard to his decisions. *This type of selection by which one tends to consider and to preserve mainly reciprocally concordant data, represents in fact, a basic component of an intuitive cognition.*

The physician's decision in the above example is based only on a probabilistic evaluation, but his overt attitude tends to appear as expressive of a sure, unquestionable interpretation. *The distance between the degree of accuracy of the information really available and the subjective evaluation of this degree of accuracy is in fact filled in, compensated for, by the specific mechanisms of intuition.*

Using a somehow different type of approach than that commonly used by researchers in overconfidence we have suggested a procedure to measure what we have termed the *intuitive acceptance of a statement* (see Fischbein, Tirosh and Melamed, 1981). We have defined the intuitive acceptance by an individual of his own solutions by resorting to two dimensions: the level of confidence and the degree of obviousness. From the fact that a person is firmly convinced that a certain statement is true, it does not follow that he considers that statement to be self-evident as well — and *vice-versa*. Therefore we have defined intuitiveness by using the formula: $I = \sqrt{C \times O}$ in which I stands for intuitiveness, C for level of confidence and O for obviousness. In the present context we only want to mention some findings related to overconfidence. It was found that for some items the majority of subjects offered erroneous answers. Nevertheless, their level of confidence with regard to these answers was higher than the level of confidence of those subjects who answered correctly. Let us consider one example.

Given a segment $AB = 1$ m. Let us suppose that another segment $BC = \frac{1}{2}$ m is added: Let us continue in the same way, adding segments of $\frac{1}{4}$ m, $\frac{1}{8}$ m etc. What will be the sum of all the segments $AB + BC + CD \ldots$ (and so on)?

One hundred and seven high school students were questioned. Six subjects answered correctly that $S = 2$. Their level of confidence (on a scale ranging from 0 to 6) was 1 (the numbers representing the levels of confidence are mean values). Fifty-five subjects answered that the sum was infinite, with a level of confidence of 2.63; and 18 subjects answered that the sum would be smaller than 2, with a level of confidence of 2.89. (The highest level of confidence obtained in that research with regard to various items was 3.5).

In the above example, the correct answer ($S = 2$) was taught in mathematics courses. *But those few pupils who remembered it were in fact very unconfident about it.* On the other hand, those pupils who gave different (but intuitively acceptable) answers were relatively highly confident about these answers. *One may then suppose that over- and under-confidence are related not only to the accuracy of knowledge of an individual (in a certain domain) but also to the degree of intuitiveness of the various items considered and the respective solutions.* The essential fact to be mentioned here is that we are very often more confident with regard to a solution never learned and incorrect but intuitively acceptable, than with regard to a learned, correct solution. In other terms: *since a thinking activity consists in a succession of decisions and since we have to believe in our decisions, our memory and our reasoning strategies tend to select and maintain statements which appear to be intrinsically credible rather than statements which have to resort to an extrinsic support (like explicit instruction).*

If a notion lacks intrinsic credibility — in the eyes of the subject — it is highly probable that he will forget it or he will not use it currently in a productive problem solving activity. In order to prevent such an effect, we tend, automatically, to resort to means which would instil in those ideas which have been imposed initially by extrinsic means (proofs, empirical findings etc.) a certain degree of intrinsic credibility. *The higher the level of intrinsic credibility of a certain notion the higher is its chance of surviving and participating actively in our solving strategies and decisions.*

It is very possible that the original idea (concept, statement, representation, etc.) will be distorted as an effect of that process without the subject being aware of the distortion.

What are the means by which intrinsic credibility is produced? Some of them have already been discussed in previous chapters (such as visualization, use of different types of models, active personal involvement etc.).

Here I would like to mention some which may be related specifically to overconfidence.

The general technique is that of producing a *coherent structure*, a Gestalt. Consequently, one tends to preserve those facts and segments which fit together and to discard those which may disturb the unity. Since the selection is not necessarily based on objective, systematic criteria, the product obtained may be intrinsically credible, but far from being correct. Usually, the selection process is organized by reference to a certain primary hypothesis or interpretation schema. A first experience, a first interpretation, may be decisive. A teacher considers that one of her pupils is a bright child because he once gave an interesting answer. The natural tendency is to stick to that primary interpretation and to continue to select and preserve those reactions of the child which may strengthen the hypothesis. Subjectively, the hypothesis is interpreted as a certitude.

The process is in fact still more complex. Facts which do not fit with the accepted schema are not necessarily altogether eliminated. They are sometimes distorted, reinterpreted, transformed in order to fit the Gestalt or at least not to disturb it. Scientists are perfectly aware of these phenomena from their own experience. Tweney, Doherty and Mynatt (1981) reviewed research indicating that: "many people, including scientists, manifest a bias to confirm. They do so by their failure to do one or more of the following:

1. Seek disconfirmatory evidence.
2. Utilize disconfirmatory evidence when it is available.
3. Test alternative hypotheses.
4. Consider whether evidence supporting a favored hypothesis supports alternative hypotheses as well." (p. 115).

They quote (p. 124) the argument of Lakatos (1978) that Popperian falsification simply does not occur: "Popper's criterion ignores the remarkable tenacity of scientific theories. Scientists have thick skins. They do not abandon a theory merely because facts contradict it. They normally either invent some rescue hypothesis to explain what they then call a mere anomaly, or, if they cannot explain the anomaly, they ignore it, and direct their attention to other problems. Note that scientists talk about anomalies, recalcitrant instances, not refutations."

Similar ineffectiveness of refutation can be observed in children's handling of mathematical problems. Consider this example:

"If a liter of juice costs 2 dollars what is the price to be paid for 0.75 l of juice?" The child has to choose the operation which solves the problem. Usually a child indicates division and he is very confident in his decision. A division operation means, in non-formal terms, cutting an object or a set of elements into equal parts. The conclusion consistent with that interpretation is that each fragment is smaller than the whole. Consequently division means "making smaller". Now, let us suggest to the child that he actually perform the division using his pocket calculator: $2 \div 0.75$. What he gets is a number bigger than 2. Very often this does not disturb the child's belief that the right operation is division. His explanation of the contradiction is that the calculator is out of order!

In other words, the finding which does not fit with the child's solution (the result is bigger than 2 and not smaller as expected) is reinterpreted. Instead of accepting that he was wrong (this would contradict his own theory about division) the child would decide that the calculator is out of order (although the probability of the second theory being correct is much smaller.)

EDUCATIONAL IMPLICATIONS: OVERCONFIDENCE AND METACOGNITION

As has been shown, there is a steady tendency of people to overestimate the validity of their own interpretations and solutions. This finding has been explained in the following way: in order to assure the continuity and fluency of our mental behavior we feel the need to rely on certain ideas which would appear to us indubitable self-consistent and self-evident. Such ideas are generally termed *intuitive cognitions*. Some of them, as an effect of their constant relevance to the reasoning activity, may get the status of axioms, postulates, principles or simply preconceived ideas. They represent the strong, the most resistent and stable component of our cognitive system.

Some intuitive ideas are correct, that is, accepted by the contemporary scientific community. But many of them are biased or totally wrong.

The very existence of these intuitive strongholds of our mental behavior creates a fundamental educational dilemma.

In the process of intellectual education the teacher has to correct his pupils' misconceptions. But very often this would imply not a simple change at the conceptual level but rather a deep reorganization of clusters of cognitive beliefs. If a pupil has claimed for instance that Lima is the capital of Columbia or that the French Revolution took place in the 19th century one may correct these mistakes by replacing some data with others. But what happens if the student has to accept that we live on a spherical, enormous and unsupported moving body; or that the operation of division can "make bigger" and that of multiplication "can make smaller"; or that the set of natural numbers and the sets of even numbers are equivalent (the same number of elements!); or that, objectively, a state of rest and a rectilinear motion with constant speed are identical, etc.? In order to accept such bizarre statements as true, the student has not merely to memorize them. He has to renounce several of his fundamental beliefs with regard to reality. He has to introduce some structural changes in his mental schemata. He has to accept (and this is the most terrifying discovery!) that, very often, *while being absolutely convinced about the truth of a certain idea, he was in fact wrong.* It is terrifying to learn that no idea is *a priori* true even if it has the appearance of self-evidence and absolutely clarity. This raises a very serious educational problem.

A child is generally not well-enough equipped with both intellectual and emotional resources to cope with this type of conflict. Let us imagine that we try to convince a child, with regard to a certain domain — let us say mathematics — that he may be wrong even when he strongly believes that he is correct. Or still more complicated: in a formal science like mathematics the fact of being wrong or correct has no absolute meaning. This may be established only with reference to a defined system of axioms. The effect may be that the very course of the student's mathematical reasoning will be disturbed. Loosening the confidence in his own intuitions (very often surprisingly wrong) he may not be able, any more, to follow systematically an idea of his own to its conclusion in order to check its validity. "How can I reach a plausible solution if at every step I may be wrong?".

Experts and, in general, adults with intellectual education have learned to play a double game. They know *in principle* that they may be wrong but they go on reasoning as if they were convinced that they are correct at every step.

The control stage may start after the entire sequence of thoughts has been accomplished.

Sometimes we are aware that we may be wrong and that we play a double game. Sometimes we forget the game completely and we are genuinely convinced that we are correct before any verification.

But, as I said, the child is not yet equipped for this kind of intellectual, very honest, duplicity. And then how can he keep on reasoning if he looses his confidence in his mathematical or scientific intuitions in general, and in his attempts related specifically to a certain problem?

Moreover, a pupil may totally lose his interest in mathematics or science, together with his confidence concerning the chosen domain.

Certainly not all the children react in the same way. Unfortunately there is no experimental evidence available so far concerning the way in which children react to intuitive conflicts, in connection with I.Q., socio-economic status, cognitive style or mathematical aptitudes. Studies in that direction would be very helpful.

Consequently we have to limit ourselves to a general discussion. It is certain that mathematics education cannot be successfully achieved by simply bypassing the intuitive obstacles through purely formal teaching.

Shall we resort to an "infiltration strategy" i.e. by smoothly infiltrating correct concepts and representations in the children's mind in the long run, hoping that the intuitive misconceptions will disappear by themselves as an effect of age and lack of use?

We may affirm now, on the basis of research and practical evidence, that intuitive biases do not disappear — neither with age nor as the effect of a lack of utilization. They remain strongly anchored in our mental schemata, they continue to be active and influential, their impact upon our thinking strategies is experienced by children and adults alike. Fundamental, spontaneous changes in the intuitive background take place no later than during the formal operational period but many intuitions become stable long before. Every instructional activity always has to cope with intuitive tendencies, and therefore contradictions between conceptual, correct structures and intuitive representations are unavoidable.

Our point of view is the following:

First, because it is not possible to bypass the intuitive obstacles, one has to cope with them on the basis of a clearly-formulated didactical strategy. One should not let the student cope by his own means with these intuitive difficulties. The potential conflict should become an active, conscious, real one by using adequate didactical techniques. I would strongly recommend that one makes clear to the child, as early as possible, that all of us, children and adults alike, encounter the same difficulties, that this is the natural, normal way in which all of us tend to think. We have to understand that in mathematics we use concepts and statements the validity of which has been established logically and not empirically. We must accept them although they sometimes contradict our natural, common sense way of thinking.

Certainly, the language used by the teacher should depend on the age of the student, on his knowledge already acquired. Historical examples would be of very great help. The fact that great scientists and mathematicians have absolutely believed in ideas that later on became obsolete, may by itself be very encouraging for the students. We have to emphasize that the fact that we make mistakes and that we usually tend to be overconfident, should not discourage us. This is normal mental behavior. What we have to do is to be aware that we may be wrong even when we are convinced that we are correct; we have to be very careful with our conclusions. At the same time, we have to learn to analyse the systematic sources of our mistakes, we have to learn procedures to check our results, and to interpret them from a broader point of view than that which would be directly suggested by the problem to solve. For instance: a problem is suggesting division as the solving operation. Does this option fit the requirements of proportionality?

Jeremy Kilpatrick has devoted an important paper to this topic especially to what has become, in recent years, known as metacognition.

Quoting Johnson-Laird he says: "People obviously know something of their own high-level capabilities — their metacognitive knowledge — but in Johnson-Laird's view they have access to only an incomplete model of their cognition. Consciousness may have emerged in evolution as a processor that moved up in the hierarchy of mind to become the operating system. The operating system has no direct access to what is below it in the hierarchy; it knows only the products of what the lower processors do. Consciousness requires a high degree of parallel processing so that the embedded mental models can be available simultaneously to the operating system" (Kilpatrick, 1985, p. 14).

Consequently we have to learn to analyse and control our own mental activity by ourselves. May we eliminate overconfidence by avoiding totally mistaken steps in a reasoning process?

In our opinion, such an aim is not achievable because it is impossible to effectuate a fluent productive reasoning activity under the permanent supervision of higher-order cognitive instances.

The analogy with a swimmer seems to be very illuminating. The swimmer cannot consciously control the details of his acts. He may only globally control the "melody" of his movements according to his purposes (with regard to speed, direction etc.). The detailed control is only automatic through direct kinesthetic and tactile feedbacks. If he consciously tries to control each detail of his movements he will certainly disrupt his swimming activity and make it practically impossible.

It is just the same with a reasoning activity. Reasoning is an automatic process guided globally by the "strategic" purposes of the thinker but only automatically in its details, through an acquired system of automatic feedbacks. Any attempt consciously to control a thinking process step by step certainly destroys its fluency, its continuity and, in fact, renders it impossible as a productive endeavor.

It does not follow that overconfidence cannot be dealt with.

An essential idea is suggested by a remark made by Robert Davis: "... good students after solving or working hard on a difficult mathematics problem, often seem to be deep in thought, turning the problem over in their minds in an after-the-fact-analysis." (Kilpatrick, 1985, p. 19).

According to Davis, an after-the-fact analysis may consolidate the students' capacity to control their thinking processes and develop metacognitive capabilities.

The problem then is not to improve calibration by merely diminishing overconfidence. By striving to convince the student that he may very often be wrong in his conception, without providing him with metacognitive techniques, we may only destroy his self confidence.

In short: developing metacognitive capabilities entails two basic aspects.

The first is that of improving the student's after-the-fact control. This would imply: (a) a global, intuitive estimation of the final result obtained and of the general strategy followed (the main line of the solving or proving strategy); (b) A detailed, retroactive analysis of the steps effectuated: checking the clarity and the correctness of the terms involved, of the processes and operations considered; making the corresponding definitions explicit, checking the theorems, the axioms, the laws, the rules (refered to implicitly or explicitly). If some mistakes have been identified it is advisable to try to determine if they were merely accidental (by lack of attention) or systematic as an effect of persistent biases in the student's interpretations or in his calculation skills.

A second aspect refers to the possibility of delegating some of these high-order conscious interventions to law-order instances, that is to say, to develop self-control schemas which would work automatically at the operating level.

There are various ways to obtain such an effect. One is to create "alarm devices" which would alert the student each time he reaches a potential pitfall in his reasoning endeavor (known from his prior experience to be such).

I once asked a young colleague, an expert in math education, how he behaves when he has to solve a potentially misleading question. (For instance: "The price of 0.65 liter of juice is 2 dollars, What would be the price of 1 liter?" The solver has to choose the right operation for solving the problem). My colleague answered that he has developed a kind of alarm system which is always triggered automatically when he encounters certain categories of problem leading to potential pitfalls. The alarm system prevents him from giving a hasty answer. In other words, it is just when the risk of overconfidence is high that the emergency brake intervenes.

A second example refers to the concept of infinity. Dina Tirosh has made an attempt to teach high school students notions related to the Cantorian theory of transfinite cardinals. One of the common difficulties was represented by the conflict between the intuitive, natural attitudes of the students and the conceptual, formally based knowledge. Naive intuitions are very often misleading in the domain of infinity. One of the main results obtained as an effect of the experimental instruction was the fact, related by Dina Tirosh in her Ph.D. thesis, that many students have developed an "alarm" technique for infinity questions. They learned to refrain from reacting according to their first impulse — usually intuitively controlled — and to look for convenient explicit theorems in order to determine the correct answer (Tirosh, 1985).

Certainly, these students have developed a better calibration in evaluating their answers. But, by diminishing the confidence in their first, spontaneous, reactions they did not at the same time lose their confidence in their mathematical capacities and their self-confidence in general. One may say, on the contrary, that the students have learned that they may acquire higher order cognitive means through which they could control their spontaneous intuitive reactions. Self-confidence was not lost. It has only changed its support. On the other hand, we suppose that the systematic use of a conscious control will also result in an improved automatic, internal control of the details of mental endeavor. We suppose that notions which are not clear, statements which are not justified, etc., may be at least partially avoided through an automatic control even before the final, higher order check intervenes. Let us emphasize again that a direct control on the micro-components of a stream of thoughts may be only automatic and it may be obtained only as an indirect effect of systematically practising a post-factum intentional analysis of the course of reasoning.

The metacognitive supervision of the course of a reasoning process may then take place either as a post factum, conscious activity or during the process itself but then only as an automatic, internal one. We are referring to a creative productive reasoning process aimed at solving a genuine problem. In this case one can not stop after each term or operation asking: Is the term clear enough to me? Have I used it according to the definition? Is my definition correct? and so on. Such a direct control would certainly destroy the continuity of the reasoning endeavor and with it the "vital impetus" as Bergson would call it — the creative stream of thoughts.

This type of external control would probably eliminate overconfidence but it would eliminate, as a side-effect, the self-confidence of the student altogether.

SUMMARY

The need for relying on apparently certain, credible representations and interpretations is, in our opinion, the main factor which explains the general tendency of people to be overconfident in their judgements.

The need for certitude leads to this type of apparently very well structured, self-consistent and apparently self-evident cognitions called intuitions.

But overconfidence is an obstacle to self-control and consequently it may block the way to a significant improvement of the quality of reasoning.

The aim of intellectual education with regard to overconfidence is not simply to obtain a better calibration of the students' evaluation of the correctness of their solutions. The aim is not merely to render the student aware of how ignorant he is or he may be. This by itself risks destroying the student's self-confidence, disrupting the very course of his reasoning and, on the other hand, does not significantly improve the quality of his mental endeavor. The genuine educational problem is to endow the student with the intellectual means: (a) to control systematically, by an after-the-fact reflection, the course and components of his reasoning processes, and (b) to develop through systematic practice some techniques for controlling directly, automatically, the validity of notions and inferences involved in the stream of reasoning.

Because intuitions and intuitive models may play an essential role in biasing notions and judgements, it is of great importance that the student should learn to identify those intuitions which may distort his representations and mislead his reactions in connection with certain areas of knowledge.

CHAPTER 4

GENERAL CHARACTERISTICS OF INTUITIVE COGNITIONS

In this chapter, the general characteristics of intuitive cognitions, some of which were introduced in Chapter 1, are discussed. The characteristics to be considered are:

1. self-evidence
2. intrinsic certainty
3. perseverance
4. coerciveness
5. theory status
6. extrapolativeness
7. globality
8. implicitness

SELF-EVIDENCE

This is the fundamental characteristic of intuitions. An intuitive cognition is self-consistent, self-justifiable or self-explanatory, (using a term of Andrea di Sessa). If we affirm that the whole is bigger than each of its parts, that every number has a successor, or that two points determine a straight line, we feel that these statements are true by themselves without the need for any justification.

Descartes mentioned evidence and certitude as the basic characteristics of intuitions (Descartes, 1967, p. 7). Spinoza gives the following example of intuition. If one has the three numbers 1, 2 and 3 one may find intuitively that the fourth proportional is 6 (6 is to 3 as 2 is to 1). The conclusion — because the numbers are small — is drawn directly, without any explicit calculation. Such a conclusion is self-evident and therefore certain (Spinoza, 1967, pp. 68—79).

Piaget has analyzed the ontogenetic construction of evidence. The main idea is that a new domain of evidence does not completely abolish the former one but integrates it as a sub-domain (Beth and Piaget, 1966, p. 195). I agree that sometimes newly acquired evidence may represent the result of an enlargement and increased flexibility of the general frame, with conservation of the former acquisitions as sub-structures (Beth and Piaget, *ibid.*). But, in general, things do not develop so smoothly. The new evidence, before becoming evidence — and even after — conflicts with the former.

Before accepting the mathematical evidence of irrational numbers and that of the larger set of real numbers, the concept of irrational numbers appeared to be in contradiction with the concept of number itself (represented by natural numbers). Moreover: two conflicting items of evidence may coexist in the same mind, and manifest themselves alternatively as such. If one is asked to compare the set of natural numbers with that of even numbers one feels on the one hand that the set of natural numbers is necessarily larger, and on the other hand that they must be equivalent (both are infinite sets).

What are the mechanisms of evidence? Why do some statements appear self-evident while others — even very familiar ones — do not? The well known formula $a^2 - b^2 = (a + b)(a - b)$ is not self-evident, while a more sophisticated idea like "every number has a successor" is self-evident.

> . . . self evidence is always without a doubt bound up with a system or with invariants common to several systems. . . . (Beth and Piaget, 1966, p. 193).

However, the above formula is organically integrated in the algebraic system of concepts and rules of calculation. Nevertheless it is not self-evident.

Self-evidence implies not only the fact that the individual is able to justify (logically or empirically) the relevant statement. An intuitively accepted statement is not only true (or apparently true); it appears to be self-explanatory or, in Kant's terminology, an *analytical judgement*. "Every number has a successor" because the concept of number *implies* the idea of unlimited iteration. "The whole is bigger than each of its parts", because the concept of a whole implies the idea of a sum of parts. $A = B$ and $B = C$ imply, *evidently*, $A = C$ because equality is intrinsically transitive. What is in A is transmitted automatically to B and so on.

Perceiving a piece of evidence means, as a matter of fact, perceiving an invariant across various potentially different manifestations. The fact that a notion is linked with a system of invariants does not ensure its self-evidence. The individual has to perceive the invariant, or the system of invariants in order to get the feeling of evidence. When one affirms with certainty that every number has a successor one perceives through a direct insight the invariant capacity of every number n to be followed by $n + 1$ etc. This is "the intuition of pure number". (Poincaré, 1920, p. 20). The statement: "Two lines parallel to a third one, are parallel to each other" is self-evident because one deals with the invariance of direction of a straight line combined with the identity of directions of the three lines. Discovering, analytically but tacitly, the invariant in the concept, one gets the feeling of the intrinsic evidence of a relation connecting the concept with an attribute or another concept.

This certainly represents a complex didactical problem. First of all, one has to decide whether it is both possible and useful to find a didactical procedure for complementing the formal understanding with a direct insight of the concept or statement. If the answer is, in principle, affirmative one has to find an adequate method for creating such a feeling for evidence in the student. For instance the theorem: "Three points determine a circle" is not evident, in contrast to the statement: "Two points determine a straight line". One may consider helping the student to see the circle statement as being an evident one. A dynamic procedure is probably helpful (a computer program may facilitate the task). One starts with one point (A): there is an infinity of possible circles. One fixes a second point. There is still an infinity of possible circles but the center moves along a determined line, the perpendicular bisector of AB. If one chooses a third point, the circle is fixed.

One may assume that, in the case of the statement "two points determine a straight line", the obviousness is generated by the subjacent behavioral evidence. One imagines one point and the infinity of possible lines passing through it. The second point fixes the line. It is an act which may be grasped in a single perception. In the case of the circle the procedure is more complicated and then one has to pass effectively from one point to two, and finally to three points, in order to grasp the successive limitations of the degrees of freedom of the circles in a single synthetic perception.

One may assume that obviousness expresses, to a certain extent, *the behavioral meaningfulness of a notion.*

To summarize, the following roots of self-evidence have been identified: behavioral direct meaningfulness; grasping of an invariant across various transformations of a given structure; internal equilibration of these transformations; the analytical character of a self-evident statement.

INTRINSIC CERTAINTY

A second fundamental characteristic of intuitive cognitions is the fact that they are accepted as certain. Self-evidence and certainty are highly correlated but they are not reducible one to the other. First of all one may be totally convinced that a statement is true without any feeling of self-evidence. We are convinced that the mathematical theorems learned in school are true but most of them are not self-evident. For instance, the theorems of Pythagoras and of Thales, the theorem referring to the sum of the angles of a triangle, are not self-evident. One accepts them only on the basis of proofs.

Certainty does not imply self-evidence. Does self-evidence imply certainty? Not absolutely. Let us consider a segment AB and on it, chosen randomly, a point C. Let us divide segment AB into two equal parts and let us again divide each of the parts into halves. Let us continue to divide in the same manner. Will we arrive at a situation such that one of the points of division will exactly hit point C?

Seven percent of the subjects answered correctly: "It depends". They got a high score on "obviousness" and a very low one on "confidence".

In other words, the subjects felt that their answer appeared *to them* as self-evident but because of the complexity of the problem they were not convinced that they were really correct.

Subjectively, the feeling of certainty and the feeling of evidence are not identical. Generally speaking when trying to identify the presence of an intuitive cognition one has to determine the extent to which it appears to the individual as an intrinsic belief. Most of our information: numerical data, names, formulas, theorems, scientific laws, are accepted because of proofs or because they are supported by the authority of a text-book or of a teacher, but they are not felt as intrinsic beliefs. They are not intuitively accepted cognitions. In contrast, the axioms of Euclidean geometry, for instance, are not only accepted because they were taught; they are also accepted as self-evident with a feeling of intrinsic certainty.

Spinoza writes, referring to the feeling of certitude:

> . . . the adequate idea of the idea *A* will be in the same mind as has the adequate idea *A*: and therefore he who has an adequate idea or . . . who knows a thing truly must, at the same time, have an adequate idea of his knowledge or a true knowledge, that is (as is self-evident) he must at the same time be certain (Spinoza, 1967, p. 70).

In other words a true idea must appear as certain because of its being true, and certitude becomes in Spinoza's view a criterion for genuine truth ". . . as light shows itself and darkness also, the truth is standard of itself and falsity" (*ibid.*).

It is clear nowadays that the feeling of certainty is not an absolute criterion of objective truth. Nevertheless, the *feeling of certainty* remains a criterion for intuitive knowledge (i.e. a criterion for a knowledge imposing itself *subjectively* to the individual as absolute). A relatively large amount of research has been devoted to the problem of confidence and to measuring the degree of confidence a person has in his own assertions. (See Echternacht, 1972; Gettys, Kelly and Peterson, 1973; Fischhoff, Slovic and Lichtenstein, 1977; Fischbein, Tirosh and Melamed, 1981; Oskamp, 1982, Lichtenstein, Fischhoff and Philips, 1982.)

The main methodological problem is to determine, by using various types of confidence test, the degree of the individual's confidence in his answers and to compare it with the objective degree of veridicality of this answer.

It has been found that people generally tend to be overconfident in their answers, that is, they are much more confident than would be warranted considering the correctness of the answers. (See Chapter 3). Most of the studies carried out so far are related to factual information and not to representations or interpretations which would qualify as intuitions. But it is remarkable that individuals tend to overlook the frailty of their knowledge and express high confidence in their solutions and interpretations even when it is not justified. (See especially Fischhoff *et al.*, 1977). As Fischhoff, Slovic and Lichtenstein have shown, people have sufficient faith in their confidence judgments (even in the case of extreme confidence) to be willing to stake money on their validity (*cf.* Fischhoff *et al.*, 1977, 559—560).

These experimental results, proving the existence of extreme overconfidence, refute the old thesis of Spinoza and Descartes according to which the feeling of certainty is in itself a criterion of validity of truth.

In a study devoted to measuring the degree of intuitiveness of mathematical statements we have defined intuitiveness according to the formula $I = \sqrt{C \times O}$ in which I stands for intuitiveness, C for confidence and O for obviousness. Confidence and obviousness have been determined by using specially adapted questionnaires (see Fischbein, Tirosh and Melamed, 1981). It has been found that in *many instances subjects display higher confidence in erroneous answers than in correct ones.*

High obviousness and high certitude with regard to a certain solution (which, combined, produce the feeling of intuitiveness) determine the robustness of the respective intuitive views. If they are not correct it is very hard to eliminate them. This category of erroneous, robust knowledge imposes a need for special didactical care. Experience has shown that robust intuitions — no matter if they are correct or not — tend to survive even when contradicted by systematic formal instruction.

PERSEVERANCE

As has just been stated, intuitions, once established, are very robust. Formal instruction which provides the student with conceptual knowledge has often very little impact on his intuitive background. Erroneous intuitions may survive together with correct, conceptual interpretations all our life. We know that we live on a spherical body — the earth — which has turned around the sun for millions of years. Nevertheless, we have great difficulty in intuitively accepting such a representation. We know that matter is composed of molecules which are composed of atoms which are in turn composed of extremely small particles moving at an enormous speed. Nevertheless, the intuitive representation of matter as being composed of moving particles is practically impossible. We agree, we have been taught, but we cannot internalize such a representation as natural and obvious. The compactness of matter — especially of solids — appears intuitively as an intrinsic property.

The survival of such contradictions between intuitive, robust representations and scientifically acquired concepts is a permanent source of difficulties for the teacher. Very often the main recommendable procedure is to make the student aware of the conflict and to help him to develop control through conceptual schemas over his intuitions.

COERCIVENESS

Intuitions exert a coercive effect on the individual's ways of reasoning. Intuitions impose themselves subjectively on the individual as absolute, unique representations or interpretations. Generally, other alternatives are excluded as unacceptable. We accept as evident that through a point outside a line only one parallel may be drawn. We cannot accept intuitively other alternatives, for instance that no parallel can be drawn (the geometry of Riemann) or that an infinity of parallels can be drawn (the geometry of Lobachevsky).

Considering the possibility of continuous functions without derivatives, the famous French mathematician Hermite wrote: "I am turning away with horror from that lamentable wound of the continuous functions without derivatives" (LeRoy, 1960, p. 329, my translation).

The following statement of Tannery is also relevant:

> I am very shocked by these points with rational abscissas, which may be as close as we want from every point of the interval (1, 0) and which, each of them, may be enclosed on a small segment, without the totality of these segments covering the whole interval (LeRoy, 1960, p. 329).

In the history of science and mathematics the coercive nature of intuitions has frequently contributed to a perpetuation of wrong interpretations and to a reluctance to accept the correct ones even after they have been logically proved. The impetus theory — a body moves because a force has been invested in it (Buridan) — prevented for a very long time the understanding of the true nature of inertia and uniform motion. The intuitive idea of the earth as center of the universe hindered the development of the correct Copernican conception of the dynamics of the solar system. I have already quoted the following example: Consider the successive divisions of a segment into equal fragments. Does one of the points of division hit a randomly chosen point on the segment? More than 77% of the subjects answered affirmatively — which is an incorrect answer. If the chosen point is an irrational one, no point of successive divisions will hit it, because these are all rational points. These pupils, starting from grade 7, have learned about irrational numbers. Despite these, they answered incorrectly. If the operation of division is infinite then, *intuitively*, it *must* reach potentially *every* point! The coercive effect of the primitive intuition of infinity prevents the student from accepting that there are points which would *never* be attained. Infinity is equivalent, in this conception, with an absolute lack of restrictions.

One cannot think about geometrical points or lines without visualizing them (using some fine graphical representation). We are trapped in these intuitive representations. We cannot think about time without spatializing it: cutting intervals of time, "long" and "short" intervals of time etc. The intuitive representation of time is either a direct feeling of duration in the Bergsonian sense — and such a representation is not manipulable conceptually — or a spatialized representation of time acceptable to both intuition and reason. In the Bergsonian interpretation of time only intuition is able to grasp pure duration. We consider that the spatialized representation of time is also a matter of intuitive elaboration. We translate directly, automatically, operations with time into spatial representations and then time becomes a matter of conceptual operation including measuring time intervals. Measuring time implies the concepts of addition, subtraction, multiplication and division of intervals. The term *interval* itself refers to both intervals of time and space. We are trapped in this intuitive representation of time as space intervals and we are not able, generally, to get rid of it. According to the Whorfian hypothesis the representation of time is a cultural product related to the structure of Standard European Languages which include the capacity to count imaginary objects (*cf.* Bolton, 1972, pp. 212—214).

A clear distinction is to be made between the coerciveness of an intuitive knowledge and the conviction inspired by a proof. If a proof is found to be wrong it is not difficult, subjectively, to renounce our conviction. But an intuitive conviction, an intuitive interpretation, cannot be eradicated easily. It forms an integral part of our mental schemata. The imperativeness of intuitions may be explained by the fact that they are not, generally, isolated mental conceptions. They express fundamental mental constraints organized in comprehensive structures. We cannot easily give up our space intuitions, because they are an integral part of our way of living and behaving. The Cantorian theorems contradict the finiteness of our mental schemas and therefore they cannot replace our natural mental attitudes.

Let me cite an example from a paper of Raymond Duval (Duval, 1983). After learning that the set of natural numbers and the set of even numbers are equivalent (together with the proof) one of Duval's subjects says:

Subject: It is strange that for such a simple question there is such a strange answer.
Exp: You are talking about the problems with pair numbers?
S: Yes.
Exp: This is a very simple question?
Subject: Yes.
Exp: Even now?
Subject: Yes, it seems to me unreal.
Exp: It seems unreal?
Subject: It is about infinity. It is strange ("C'est à l'infini, c'est bizarre").
Subject: In the set of natural numbers there are always the even numbers *and* the odd numbers. And one has said that the two sets are equivalent ("Et l'on a dit qu il y a *autant* de nombres entiers que de nombres pairs").
Exp: Are you sure of your answer?
Subject: Oh yes! (Duval, 1983, pp. 397—398, my translation).

And Duval comments:

Thierry (Subject 1) did not become aware of the paradox merely as an effect of recognizing the objective contradiction between two opposite conclusions as he has expressed them at the end of his last intervention. He became aware of the conflict only because one of the conclusions represented an evidence which was so strong, so striking that it seemed to exclude any possibility of interrogation" (Duval, 1983, p. 400).

Becoming aware of a contradiction is different from a conflict between two view points as they may emerge during a discussion: in a discussion it is a part of the game to admit the relativity of confronted positions as long as one has not overruled the other (Duval, *ibid.*).

It is a basic difference between the relativity of a, somehow, conventional viewpoint — as exposed in a formal-logical dispute — and the apparent absoluteness of an intuitive acceptance. The second is relevant for the coercive nature of intuitions.

THEORY STATUS

An intuition is a theory (or a mini-theory) never a mere skill or the mere perception of a given fact.

Accepting intuitively that "through a point external to a line one may draw one and only one perpendicular to that line" one accepts the necessity and generality of that statement.

We affirm that "two intersecting lines determine pairs of equal opposite angles" and we claim that this is self-evident. Certainly, perceiving the image, we see the equality of the angles. But this is a perception not an intuition. *What we intuit is the universality of the property.*

The theoretical property of intuitions entails several aspects. An intuition is never confined only to stating the universality of a property nor to the perception of a certain fact. In an intuition one generally grasps the universality of a principle, of a relation, of a law — of an invariant — through a particular reality. Accepting intuitively the postulate of Euclid does not mean that *practically* we are able to draw a certain segment line parallel to another segment line. Accepting intuitively the postulate means that we feel that — *beyond any practically possible evidence* — we are absolutely convinced that the two lines may be extended indefinitely without crossing each other, in both directions.

At the same time, we imagine the two parallel lines as potentially infinite extensions of the particular segments. The line segments confer intuitiveness on the general property expressed in the Euclidean postulate. Mentally continuing the lines in both directions we "see" dynamically that they may be indefinitely extended without crossing one another.

An intuition, then, is not a pure theory. An intuition is a theory expressed in a particular representation using a model: a paradigm, an analogy, a diagram, a behavioral construct etc.

EXTRAPOLATIVENESS

This brings us to the property of extrapolativeness. "It appears", writes Westcott, "that intuition can be said to occur when an individual reaches a conclusion on the basis of less explicit information that is ordinarily required to reach that conclusion" (Westcott, 1968, p. 97).

Westcott has used this definition as the theoretical basis of a method of measuring intuition. In this, the subject had to solve five clue problems. He had to determine the final clue on the basis of information obtained successively, by request, about the previous clues. For instance, the subject sees: "January". He asks for more information (a second clue). He gets: "February". He may then decide that the fifth clue should be June. Another example: the first clue is: 326—1957. Asking for a second clue the subject gets 732—6195. He cannot solve the problem. He asks for a third clue, which is 573—2619. He then decides that the fifth should be 195—7326 (Westcott, 1968, p. 215). The measure of the intuitive capacity of the individual depends on how many clues he needs in order to solve the problem. An individual with a high intuitive capacity would guess the solution relying only on a small number of cues. Interestingly, Bartlett suggested much the same technique as a possible measure of intelligence, when he speculated "whether there may be a direct relation between capacity to utilize minimal information (in terms of number of items) . . . and high ranking intelligence" (Bartlett, 1958, p. 31).

An intuition always exceeds the data on hand. However being an extrapolative guess is not sufficient to define an intuition. A feeling of certainty is also a necessary characteristic of an intuition. Otherwise *it is a mere guess.* It is this particular combination of incompleteness of information and intrinsic certitude which best characterizes an intuition.

The extrapolativity aspect is not always evident, because the apparent obviousness of intuitions hides the incompleteness of the information on which they are based. For instance, consider again the postulate: "Through a given point not on a straight line, one, and only one parallel line can be drawn". The statement is intuitively evident; it does not seem to require additional information, nor any logical proof. In fact, the statement extrapolates a very limited experience to infinity. But the apparent obviousness of the statement hides the need (and, of course, the impossibility) of further proof.

Poincaré has emphasized the intuitive leap which characterizes the recurrent method of reasoning. If a statement is true for 1 and if true for $n - 1$, it is also true for n, one concludes that the statement is true for every n. The reasoning proceeds through "a cascade of hypothetical syllogisms". If true for 1 than it is also true for 2; if true for 2 than it is also true for 3, etc. It is enough to conclude with absolute certitude that *the theorem is true for every n.*

The principle on which mathematical induction relies is not gained by experience. Through experience one may learn that the inference under discussion is true for the first ten or for even the first hundred numbers but not for the whole infinite set of natural numbers.

On the other hand there is no analytical proof for supporting the recurrent type of inference. *When dealing with infinity, experience and analytical reasoning are not able to yield, by themselves, a basis for absolute confidence.* It is clear also that mathematical induction is not based on conventions, as some geometrical axioms are. On what, then, does the universality of a conclusion drawn on the basis of a recurrent proof rely?

Why (asks Poincaré) does this reasoning impose itself on us with an irresistable evidence? It is because (he says) it is the affirmation of the power of the spirit which feels itself capable of conceiving the infinite repetition of the same act if this act has been once possible.

The spirit possesses a direct intuition of this power and the experience represents only an opportunity to use it and to become aware of it (Poincaré, 1906, pp. 23—24).

In LeRoy's concise version: "The mere intuition produces a logical infinite" (LeRoy, 1960, p. 337).

This kind of intuitive leap has to intervene always when dealing in mathematics with infinite processes or infinite sets. As long as one has to do with the dynamic form of infinity there is no apparent difficulty. It seems that one is naturally able to conceive of the indefinite continuation of a process like that of constructing always greater numbers or of extending indefinitely a line. It is this type of situation that Poincaré has referred to when mentioning "the power of the spirit which feels itself capable of conceiving of an infinite repetition of the same act . . .". *The notion of dynamic infinity expresses directly, in the purest way, the extrapolative capacity of intuition itself.*

Things change radically when one tries to pass to actual infinity.

David Tall has shown that many of his students, when asked about the meaning of 0.9999 . . . , do not accept that 0.999 . . . = 1. One may intuitively conceive the unlimited sequence 0.9, 0.99, 0.999 etc., but there is a fundamental difficulty in seeing that 0.9 = 1 simply because this would imply that the *infinite succession has been actually realized.* It is the same when comparing infinite sets. The Cantorian theorems are generally unacceptable intuitively. From this fact *one may learn that the extrapolative capacity of intuition does not apply to actually given infinite sets.* The extrapolative capacity of intuition is dynamic, constructive in its very nature. Its predictive and explanatory capacity ceases if one refers to infinite sets considered as actually given. Intuitively, such sets would mean the final state of an endless process, *which is contradictory in natural, intuitive terms.*

One may then assume that, psychologically, the universality grasped by intuition through a particular, given instance does not refer to an actual universality but rather to a potential one. When, on the basis of induction we tend to generalize intuitively, we mean that the concept (the universal idea) is a predicate which one may attribute *potentially* to a class of elements. When we affirm that the shortest way from a point external to a line to that line is a perpendicular drawn from the point, we do not mean that we have tried and compared actually all the distances. We only mean that *it is possible* to try indefinitely and that we will always get distances which will be longer than the perpendicular. An actually given infinity is a pure logical, conceptual construct, not intuitively acceptable — it has no behavioral meaning, it cannot be self-evident and intrinsically coercive.

As I said, an intuition is a theory, not the assertion of a certain given fact. By virtue of its specific extrapolativeness intuition exceeds the facts afforded by the information at hand. But that conjectural structure of intuitions is generally hidden by their apparent obviousness and certainty. This is in fact the fundamental role of intuitive cognitions: to confer certainty on extrapolated ideas.

GLOBALITY

Intuition is also described as a global, synthetic view, as opposed to analytical thinking which is discursive in its very nature. LeRoy writes in connection with intuition:

> It is always a direct, quick vision, a synthetic view without any preliminary analysis. It is clear that the discourse which progresses step-wise, which connects the notions one by one, which goes slowly from one point to another, would not be sufficient for establishing the science. This process is constantly joined by a different demarche in order to orient it, to animate it: the lightning of a sudden illumination, the condensed lecture of a vast virtual ensemble in a reduced image. (LeRoy, 1960, pp. 327—328, my translation.)

The global character of intuition is reminiscent of the concept of Gestalt. Intuition is a structured cognition which offers a unitary, global view (or insight) of a certain situation. One may consider that the laws governing the crystalization of intuitions are partially derived from, or at least similar to, the laws of Gestalt. For instance, one may plausibly connect the role of analogy in structuring an intuitive view with the fact that the meaning of a Gestalt is determined by its basic internal dependencies rather than by the discrete elements from which it is composed.

If one looks at black-and-white photographs of a person, one recognizes the person immediately, despite the fact that the absolute values of colors and sizes are different. One identifies the image by grasping the Gestalt, not by considering the details.

If a pupil is, for the first time, confronted with the problem of finding the formula for the volume of a prism, this formula may be inspired by analogy with the formula for calculating the area of a rectangle. *Globally*, the two situations are similar, there is the same basic idea in both procedures — one multiplies the base by the height. The transfer from one situation to the other is not made through deduction but rather by grasping intuitively, directly, the common global situation and the idea of the solution in both cases.

Michael Polanyi has given extensive attention to the role of Gestalt in intuitive and, in general, in scientific knowledge.

He writes:

> ... in the structure of tacit knowledge we have found a mechanism which can produce discoveries by steps we cannot specify. This mechanism may then account for scientific

intuition — such intuition is not the supreme immediate knowledge, called intuition by Leibniz or Spinoza or Husserl, but a work-a-day skill for scientific guessing with a chance of guessing right (Polanyi, 1969, pp. 143—144).

How does scientific intuition work (and in fact intuition in general)?

Polanyi sees a deep analogy between the integrative capacity of intuition and that of perception:

... Where to turn for a logic by which such tacit powers can achieve and uphold true conclusions? *We must turn to the example of perception.* Scientific knowing consists in discerning Gestalten that indicate a true coherence in nature. (Polanyi, 1967, p. 137.)

Such an integrative, tacit process is based on two types of clue: subliminal ones and marginal ones. The subliminal clues are totally unobservable while the marginal ones are observable from the corner of the eye, but generally they are not in the focus of our attention. Both subliminal and marginal clues are subsidiary to the focal awareness of the object (Polanyi, 1967, 139—140).

Intuitions may be more or less structured (internally organized) and, as a consequence, more or less stable. There are incipient immature intuitions which may disintegrate relatively easily under the impact of some conflicting, striking opposite evidence. A teacher observing the behavior of one of his pupils in a new class may assume intuitively at the beginning that this child is a hard-working, well-motivated learner. As a matter of fact the teacher may discover later on that the good impression was based only on superficial, hasty information. There are, on the other hand, strong intuitions deeply rooted in the experience of a person, very well articulated internally and, at the same time, very well articulated with the entire structure of the person's mental skills and schemata.

The aggregate of space intuitions developed in early childhood is an example of such a perfectly structured intuitive system.

The globability of intuitions — based on tacit elaborations — is generally expressed in a selection process which tends to eliminate the discordant clues and to organize the others so as to present an unitary, compact meaning. It is this type of coherence which explains also the resistance to change, the robustness of established intuitions.

IMPLICITNESS

It has been affirmed here that intuitions appear, generally, to the individual as self-evident, self-consistent cognitions. This does not exclude the assumption that the intuitive reactions are in fact the surface structure expression of tacit, subjacent processes and mechanisms.

The phenomenon of extrapolation, to which we have referred above, usually acts in an unconscious manner. The student who accepts the Euclidean postulate of the parallels is generally not aware that he extrapolates a limited experience beyond any possibility of verification. A person asked to define the notion of solids is generally not aware of the fact that he has in mind the paradigm of a certain solid object — for instance a piece of metal or a stone. The definition is inspired by the particular instance. Confronted with the task of allocating flour to a certain category (solid, liquid) the subject is puzzled.

A person claims that after getting tails several times in succession "the chance is higher to get a head" (the negative recency effect). He is not aware of the mechanism of his wrong (intuitive) prediction. The chance of getting the sequence TTTT is exactly the same as that of getting the sequence TTTH *in the above order*. But the chance of getting T three times and H once is higher than that of getting T four times (no matter the order). The two problems are confused unconsciously in the subject's mind.

Anderson and Wilkening claim that when intuitively comparing various magnitudes (for instance areas, speeds, chances) a certain algebraic calculation (addition, multiplication etc.) takes place unconsciously — only a detailed statistical analysis may reveal these tacit operations (Wilkening and Anderson, 1982, 1984).

A child who "conserves", compensates (unconsciously) the variations of dimensions.

An individual who draws the trajectory of a falling body, dropped while it was in motion, as a "straight down" line, is unconsciously influenced by his impetus theory (after being dropped the body, not possessing its own horizontal impetus, follows only the trajectory imposed by its weight).

According to Jung, intuition perceives, unconsciously and uncritically, possibilities, principles, implications and situations as a whole (*cf.* Westcott, 1968, p. 34).

The tacit character of the processes on which an intuition is based explains its apparent obviousness. It makes self-control over intuitions a difficult task. It also considerably complicates the work of the researcher. Not only does intuition hide its tacit strategies, it is automatically opposed to any analysis since this would annihilate its intrinsic certainty, its compactness, its robustness. As a result of such an analysis the individual risks getting confused in his reasoning activity.

In certain instances, intuitive evaluations or interpretations use deliberately intuitive means — mainly intuitive models and practical activities. For instance, the solar system analogy is used for interpreting the structure of the atom. A tree diagram is used for combinatorial problems. Oriented line segments are used for representing vectorial magnitudes. But in such cases, too, the intimate processes by which the intuitive understanding arises are mostly unknown to the individual.

The identification of the original with the model, and *vice versa*, takes place on the ground of certain rules of which the individual is not aware. The mapping processes are usually performed globally from Gestalt to Gestalt.

SUMMARY

Trying to summarize the various features described so far of an intuitive cognition, one may feel somehow puzzled. These features may appear contradictory or at least disconnected. While immediacy may be considered to be related to globality, what have these two aspects to do with an extrapolative inference? How does it happen that the feeling of intrinsic conviction is associated with what is no more than a plausible guess?

The picture seems to become meaningful and consistent if one resorts to the metaphor of sensorial perception.

If you prepare to cross the street, you look around to observe approaching cars. In a global, quick view you see the cars, you evaluate their speed, you perceive the decreasing distance separating them from you. You obtain a global, but perfectly structured picture, not only of what has already happened, not only of what is happening at a certain moment, but also of what is expected to happen in the near future. As a consequence, you adapt your behavior in an almost reflex manner to all this inflowing information. There are many things you do not know. You do not know, for instance, the exact speed of the vehicles. You don't know with certitude what will happen in the next moment, what the drivers' intentions are.

However, you mix all these past, present and future assumed events in a global image which almost automatically dictates your immediate behavior.

The analogy with an intuitive cognition is striking. In an intuitive cognition, the "given" and the "plausible" are mixed in one global, apparently self-consistent, self-evident, certain idea, which inspires and guides the strategy of the next mental steps. Polanyi, as I have already said, has also used this analogy between intuition and perception.

But our theory is that one has here more than a simple metaphor. Intuition fulfills, at the intellectual level, the function fulfilled by perception at the sensorial level: intuition is the direct, cognitive prelude to action (mental or practical). It organizes information in a behaviorally meaningful and intrinsically credible structure.

The next chapter continues the analysis of intuitions by proposing various classification schemes.

CHAPTER 5

THE CLASSIFICATION OF INTUITIONS

In order to introduce some clarity into the complex domain of intuition various classification attempts have been made.

Henri Poincaré described (a) intuitions related to the senses and imagination; (b) intuitions expressed in empirical induction, and (c) the intuition of the pure number, which represents the source of mathematical induction (and, generally, of mathematical reasoning) (Poincaré, 1920, p. 20).

More recently, Bahm (1960) has mentioned three types of intuition: objective intuition (immediate apprehension of the external world); subjective intuition (immediate apprehension of the self) and organic intuition (in which the object and the subject appear immediately together in apprehension) (Westcott, 1968, p. 19).

PIAGET'S CLASSIFICATION

The classification by Piaget is much more complex. He mentions several possible dichotomies. A first dichotomy distinguishes *empirical* and *operational* intuitions. Empirical intuitions refer to the evaluation of physical properties of objects (for instance, the weight of an object), or to real psychological experiences known by introspection (for instance, the intuition of duration). Operational intuitions refer to actions related to objects and psychological phenomena.

A second dichotomy, applied only to operational intuitions, distinguishes intuitions which are accompanied by images (geometrical intuitions) and intuitions which do not possess this property (operations with discrete objects). Geometrical intuitions are characterized by the fact that they are homogeneous to the respective logical operation (Beth and Piaget, 1966, pp. 223—225). Let us consider, for instance, the notion of the circle. The *image* of the circle (as a geometrical figure) and the *concept* of it are completely congruent. (There is nothing more in the circle, as a geometrical figure, than is expressed in the given concept).

Things are different when one performs, for instance, a certain classification. The images of the objects on which the operation of classification is performed and the operation itself (as a logical process) remain essentially heterogeneous. Only when dealing with space is there a complete isomorphism possible between the concept and the corresponding image.

A second dichotomy referring to operational intuitions suggests the more general distinction between *pictorial* intuitions in general, that is, intuitions expressed by images (intuitions imagées), and operational intuitions in a strict sense (i.e. intuitions referring to logico-mathematical concepts) (Beth and Piaget, 1966, p. 224).

It is difficult to follow Piaget's classification because of the generality he confers on the term intuition. In Piaget's terminology an intellectual activity is either intuitive or formal. Consequently, almost every intellectual activity the child is able to perform before the formal operational period may be considered as being achieved on an intuitive basis.

CLASSIFICATIONS OF INTUITIONS BASED ON ROLES AND ORIGINS

By contrast with Piaget's analysis, the classifications of intuitions to be proposed here are related to their roles, their origins, their relationships to other types of cognition.

Let me say from the beginning that these distinctions are far from being absolute. As we have frequently emphasized, intuitive forms of knowledge, although very different in their specific manifestations, reflect nevertheless the same basic, adaptive mechanism.

Classification Based on Roles

According to a first classification intuitions may be grouped into *affirmatory, conjectural, anticipatory* and *conclusive* intuitions.

This classification considers the relationship between intuitions and solutions. In affirmatory intuitions the solution element is implicit. In conjectural intuitions the solution aspect is explicit but not involved explicitly in a solving endeavor. In anticipatory intuitions both the solution moment and the problem solving framework are explicit. In conclusive intuitions, the individual is already beyond the analytical search effort which has followed the initial anticipatory intuitive "flash", and the solution appears intuitively closed and intrinsically directly acceptable.

Affirmatory intuitions are representations or interpretations of various facts accepted as certain, self-evident and self-consistent. For instance: "two points determine a straight line "or" the whole is bigger than each of its parts". Affirmatory intuitions may be subdivided according to two different criteria.

One subdivision classifies affirmatory intuitions into *semantic, relational,* and *inferential. Semantic* intuitions are those referring to the meaning of concepts. For instance, the concept of a straight line has, in geometry, a non-intuitive axiomatic meaning according to a given system of axioms. But it has also several related intuitive meanings: a behavioral one (the shortest way between two points); a physical one (a light beam); a graphical one; a functional—material one (a string very well stretched). A point may be undefined (in an axiomatic presentation), axiomatically related to lines and planes. But it also has an intuitive correspondent — a point represented by a small spot or a small piece of matter "as small as we want". The notion of force has an axiomatic meaning determined by its relation with mass and acceleration ($f = ma$). But it is also connected with various intuitive — basically behavioral — interpretations like the feeling of effort or the notion of "a force as a mover".

A *relational, affirmatory intuition* is expressed in apparently self-evident, self-consistent statements. Let me mention some examples. "Through a point outside a line one may draw one and only one parallel to that line" (the famous fifth postulate of Euclid). "The whole is bigger than each of its parts". "In order to maintain the motion of a body, a certain force is needed" (the impetus theory). "A heavier object falls faster than a lighter one". Some of these statements are correct (in the realm of a certain axiomatic system). Others are incorrect but we tend naturally to accept all of them as self-evident and certain.

Inferential affirmatory intuitions may have an *inductive* or a *deductive* structure. After one has found that a certain number of elements (objects, substances, individuals, mathematical entities etc.) have certain properties in common one tends *intuitively* to generalize and to affirm that the *whole* category of elements possesses that property. This is not a mere logical operation. The generalization appears more or less suddenly with a feeling of confidence. This is a fundamental source of hypotheses in science. According to Poincaré "generalization by induction, copied, so to speak from the procedures of experimental sciences" is one of the basic categories of intuitions (Poincaré, 1920, p. 20).

One may also describe deductive forms of inferential intuitions. For instance, from $A = B$ and $B = C$ one deduces *directly* as a self-evident conclusion that $A = C$. Or take the following example of intuition mentioned by Poincaré: "If on a straight (line) the point C is between A and B and the point D between A and C then the point D will be between A and B" (Poincaré, 1920, p. 19). The conclusion is accepted intuitively "as an appeal to imagination" (*ibid.*). The last examples are not mere concepts or statements. They are logical inferences but, nevertheless, the relation between the premises and the conclusion is accepted as self-evident, as intrinsically necessary.

Affirmatory intuitions may also be classified according to a different criterion into *ground* and *individual* intuitions. We call ground intuitions all those basic representations and interpretations which develop naturally in a person — generally during his childhood — and are shared by all the members of a certain culture. Space and time representations, intuitions related to causality, to basic physical properties, etc., belong to this category. The three-dimensional representation of space, the intuition of duration ("le temps veçu" of Piaget which gets a metric structure when spatially expressed), the idea that every event must have a cause, the automatic adaptation of predictions to the objective frequency of certain events

(Fischbein, 1975, p. 20—65), all of these are examples of ground (common, basic) intuitions.

On the other hand, individuals acquire personal intuitive representations related to their life and activity. "I do not believe in John's promises, my intuition tells me that he is a liar"; "Americans are naive people"; "I am not a professional psychologist, but my feeling is that I.Q. tests are very often misleading tools".

In brief: affirmatory intuitions may be classified on the one hand into semantic, relational and inferential (inductive and deductive); and on the other hand into ground (common) and individual intuitions.

In the case of affirmatory intuitions one *affirms*, one claims something. *Conjectural intuitions* express by their very nature an assumption about future events, about the course of a certain phenomenon etc. Such a conjecture is an intuition only if it is associated with a feeling of confidence.

There may be either lay or expert conjectural intuitions: "This child will become a brilliant mathematician"; "I will invest my money in that business, I am sure that it will be successful"; "From what John has told me I am sure that he will soon leave the country". These are examples of lay intuitions. They are not based on special expertise but rather on everyday experience.

Professionals, people with a rich experience in a certain domain, develop particular expert intuitions connected with their domain of activity. Doctors, teachers, engineers etc. are able very often to take decisions in their domain only on the ground of an apparently minimal amount of information which they are able to use with high perspicacity. The specialist is able to select the information obtained so as to grasp the most relevant aspects of it, to determine its significance, to weigh the probability of various possible interpretations in the given circumstances and to organize the whole in a meaningful highly plausible conclusion. What is fundamentally important is that the expert has the capacity to convert into relevant messages apparently obscure, non-salient aspects of the situation. All of this may be done automatically before any systematic, complete analysis is made and the result appears then to be an intuitive, global evaluation.

Berne (quoted by Westcott (1968)), published a series of papers from 1949 to 1962 in which he develops a psychodynamic view of intuition based on his personal experience as a psychiatrist at a military selection centre. He had from 40 to 90 seconds to evaluate each of the approximately 10 000 men he had to examine on the basis of the questions: "Are you nervous?", and "Have you ever been to a psychiatrist?" In order to introduce some variety in his activity Berne began to predict the civilian occupations of the men, their personal history etc. on the sole basis of a global evaluation of the behavior and the appearance of the examinees. Berne affirms that in these conjectural evaluations, an essential role was played by the "noise" in the interpersonal communication system. It is mainly the noise component which is relevant for the state of a system. To the repairman of a TV set what is important is not the content of the image on screen but the "noise" produced, for instance the snow perceived on the screen. It is the same when evaluating an individual intuitively. It is not so much what he says but how he says it, in what circumstances. His behavior, his attitudes etc. may be decisive. Conjectural intuitions are certainly of fundamental importance in every professional activity.

Anticipatory and *conclusive* intuitions represent the third and fourth categories in the classification in which affirmatory and conjectural were the first two. They may be grouped together as *problem-solving intuitions.*

Anticipatory intuitions represent the preliminary, global view which precedes the analytical, fully developed solution to a problem. This type of intuition is frequently mentioned, using various terms, by those working in the domain of problem-solving. What distinguishes affirmatory and anticipatory intuitions is their respective role in a cognitive endeavor. Through an affirmatory intuition one accepts as self-evident a certain notion, a certain statement — the notion of a straight line, the fifth postulate of Euclid etc. An anticipatory intuition does not simply establish an (apparently) given fact. It appears as a discovery, as a solution to a problem and the (apparently) sudden result of a previous solving endeavor.

On the other hand what distinguishes conjectural and anticipatory intuitions in our classification is that anticipatory intuitions represent *a phase* in the process of solving a problem (necessarily followed by an analytical endeavor), while conjectural intuitions are, more or less, *ad hoc* evaluations and predictions generally not included in a systematic solving activity. Obviously the distinction is not, and cannot be, absolute. In fact, we have to consider a continuum from affirmatory to anticipatory intuitions passing through conjectural ones. In all of these categories both components are present: the affirmatory attitude and the element of conjecture. It is the situational context which is decisive. Certainly there are also intrinsic differences. The solution attitude is relatively tacit in affirmatory intuitions while it is explicit in conjectural and anticipatory intuitions.

In anticipatory intuitions there is generally a certain need (not subjectively felt) for an analytical control. The confrontation between the intrinsic certitude and the externally imposed demand for verification may give rise to very interesting psychological (sometimes conflictual) situations.

Let us consider in more detail the problem of certitude in conjectural and anticipatory intuitions. An affirmatory intuition appears as self-evident and intrinsically certain to the individual. May we claim the same about the other two categories? Objectively they are only plausible guesses, conjectures. Are they true intuitions?

Is it possible that a conjecture, a preliminary hypothetical solution, appears to the solver with a high degree of certainty as it does in affirmatory intuitions? Would it not be reasonable to affirm that what we have called conjectural and anticipatory intuitions are not genuine intuitions according to our definition? Expressing only conjectures, it seems that they could not be certain.

The answer to this apparent dilemma may be found in an important psychological phenomenon generally overlooked. There are in fact *two* types of conjectures. There are conjectures deliberately, formally produced while considering the various possible relations among variables in a certain situation. It is to this type of conjectures to which Piaget is referring when describing the general properties of the formal operational period. It is said that an adolescent is able to identify *hypothetically* the various factors which may affect a certain phenomenon and to set up, accordingly, a systematic list of hypotheses. This type of assumption does not represent *intuitions*. The whole process takes place in a deliberate, explicit, fully controlled manner. There is also a second type of conjecture, which may appear as conjectures (hypothetical, uncertain) only *after* some objective analysis has been carried out. *During* the solving endeavor itself, they may appear, *subjectively*, as moments of illumination, as certain, evident, definitive, globally grasped truths. These are anticipatory intuitions.

Poincaré, in his autobiographical note concerning invention in mathematics, mentioned the following episode:

> One morning walking on the bluff, the idea came to me with just the same characteristics of brevity, suddenness and immediate certainty that the arithmetical transformations of indeterminate ternary indefinite forms were identical to those of the non-Euclidean geometry (Poincaré, 1913, p. 388).

This was not a conjecture deliberately and formally produced as such, while considering the various theoretical possibilities in a given situation. The author did not consider it, at the very moment of the discovery, as a conjecture. It appeared to him not as a plausible guess but as an *absolute truth*. As Poincaré tells in continuation, *it was only afterwards* that he started to analyse this result and to draw a number of consequences. Cantor related a similar story about his discovery of the equivalence between sets of points belonging to figures with different numbers of dimensions.

Let me add a quotation from a paper written by David Tall, a young mathematician and a brilliant psychological analyst.

Tall tells the story of a discovery he made related to what has been called "non-standard analysis".

Referring to the moments of "insight" he got during "the tortuous route" by which he came to build the theory, Tall writes:

> A classic description of "problem solving" involves "conjectures" which are then checked out. Here the researcher never felt that he made "conjectures"; what he saw were "truths" evidenced by strong resonances in his mind. Even though they often later proved to be false, at the time he felt much emotion vested in their truth. There were no coldly considered possibilities. They were intense intuitive *certainties*. Yet at the time his contact with them often seemed tenuous and transient; initially he had to write them down, even though they might be imperfect, before they vanished like ghosts in the night (Tall, 1980, p. 33).

How far this description is from the "cold" process of setting up lists of formally envisaged possibilities as assumed by Piaget when referring to the formal operational period!

The fundamentally contradictory nature of intuition is splendidly suggested by Tall's introspective note.

The new ideas appeared as "intense, intuitive certainties" and at the same time as "tenuous and transient"! Such a contradiction would be impossible at an analytical, conceptual level. Formally, something which is "imperfect" cannot be accepted as certainly true! Yet at the intuitive level, this situation takes place. It is a strange situation, considered from an analytical point of view. But it is exactly the role of "intuition" to confer on an imperfect (incomplete, even erroneous) solution, the appearance of perfection and of intrinsic certitude.

An anticipatory intuition is a preliminary solution to a specific problem while an affirmatory intuition represents a stable cognitive attitude with regard to a more general, common situation.

But, again, a clear-cut distinction cannot be made. Moreover, one may assume that anticipatory intuitions are inspired, directed, stimulated or blocked by existing affirmatory intuitions. A plausible conjecture is the product of a constellation of intuitive blockages and drainages.

If one is asked: "How much is 3/4 of 120 cm?" one does not look, generally, for the solution by multiplying, $120 \times 3/4$, but one tends naturally to determine first one fourth ($120 \div 4$) and then one multiplies the result by 3. One tends intuitively to avoid multiplications in which the operator is not a whole number.

Children, asked about the price of 1 liter of wine if 0.75 liter costs $2 are looking naturally for a solution by multiplication because, intuitively, multiplication means "making bigger".

For hundreds of years mathematicians tried to find a proof for the fifth postulate of Euclid. The postulate seemed certain but not self-evident enough. The basic problem with this postulate — no matter in which form it is expressed — is that it admits, implicitly or explicitly, the possibility of infinite straight lines. The mathematicians then became aware of the fact that the postulate needs some kind of elaboration: either to find a more self-evident, equivalent statement or to try to deduce it from the other nine axioms of Euclidean geometry. The form presently used in school textbooks is that proposed by John Playfair in 1795 (Kline, 1980, p. 79).

As is very well known, none of these attempts succeeded and non-Euclidean geometries became a part of modern mathematical thought.

What I am trying to emphasize is that the intuitive acceptance of the fifth Euclidean postulate was so strong that it inspired two thousand years of research in a wrong direction! It took 2000 years of unsuccessful efforts until mathematicians dared to consider some intuitively incredible alternatives!

The conclusion which follows from the above is that the illuminatory moments — called by us anticipatory intuitions — are not unequivocally and specifically dictated by the data of a given problem. They depend in a major degree on the individual's own robust, stable intuitive attitudes. To put it differently, in order to predict a person's strategy in solving a problem, his assumptions and his ways of verification, it is not enough to know what his formal knowledge in the respective field is and the data of the given problem. One has to have some reliable information about the "lines of force" of his intuitive, affirmatory, aggregates.

Conclusive intuitions summarize in a global, structured vision the basic ideas of the solution to a problem, previously elaborated. As Hadamard has written:

> . . . any mathematical argument, however complicated, must appear to me as a unique thing. I do not feel I have understood it as long as I do not succeed in grasping it in one global idea . . ." (Hadamard, 1949, p. 65). Later, he refers to: ". . . that synthesis given the leading thread, without which one would be like the blind man who can walk but would never know in what direction to go. (Hadamard, *ibid*., p. 105.)

This first classification may be schematized as follows:

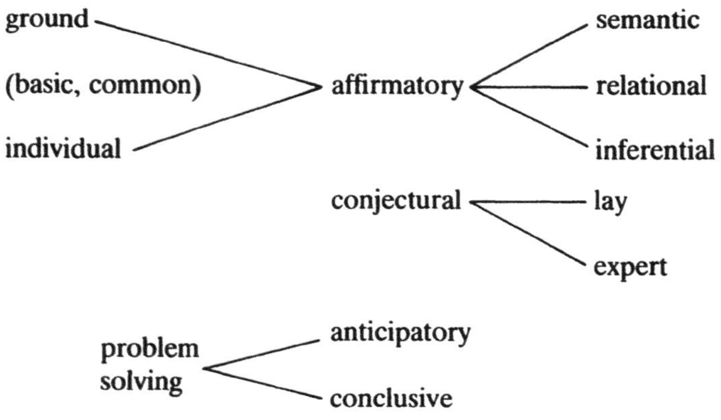

Classification Based on Origins

A second basic classification refers to the origin of intuitions and concerns mainly affirmatory intuitions. According to this criterion one may distinguish *primary* and *secondary* intuitions. Primary intuitions, in turn, may be either pre-operational or operational.

Primary intuitions refer to those cognitive beliefs which develop in individuals independently of any systematic instruction as an effect of their personal experience. Primary intuitions, then, include both ground (general, common) intuitions and individual ones (produced by particular but natural, normal circumstances).

All the examples mentioned above with regard to ground intuitions are generally primary intuitions. In addition, a child may develop special intuitions, for instance for distinguishing colors, for evaluating distances or weights, for his orientation in space etc. according to his usual way of life and activity.

Primary intuitions may be either pre-operational or operational. Let us consider some examples of pre-operational intuitions. A five-year-old child considers intuitively that by altering the form of a piece of clay one alters also the quantity or the weight of it. A child may consider intuitively that we can make the clouds move by walking (the magical stage of causality — *cf.* Piaget, 1930). A five year old child would apprehend intuitively the number of elements in a row according to the length of the row and not according to bijection criteria. An example given by Piaget and Szeminska is the following:

Boq (4; 7): Exp: — Put here the same amount of sweets as there. This (6 sweets) is for Roger. Now take for yourself the same amount. (The child puts 12 sweets in row, so close to each other that the row obtained was shorter than the previous one. Is it now the same? *Not yet* (the child adds sweets). And now? *Yes.* Why? *Because it is so* (he indicates the length of the rows) - - - - At the end of the interview Boq is offered two rows of sweets, one containing 3 and the other 4 sweets more closely packed than the first; the 3 row was longer: Where are more? *Here* (the 3) Why? *This line is longer* (Piaget and Szeminska, 1964, p. 99).

These are examples of pre-operational intuitions. The appreciation is intuitive, global, without hesitation, based on configurations rather than on operational criteria. The length criterion is made explicit only by request.

After the age of 6—7 new intuitions develop based on the composability and reversibility of intellectual operations: intuitions related to conservation capacities, to the notions of number and cardinality, to elementary logical and arithmetical operations. A concrete-operational child understands and uses intuitively class relationships, bijections, correspondences, notions and operations related to order and seriation (for instance the transitivity of the relation of order). His causality ideas become of a mechanical type. His interpretations of reality liberate themselves from the egocentric perspective: Gut ($9\frac{1}{2}$): "Why do the clouds move more or less quickly? *Because of the wind. They move along by the wind*" (Piaget, 1930, p. 72).

These are examples of operational intuitions. They are no longer based on configurations but rather on operations (in the Piagetian sense). In many situations, the reactions of the child remain, nevertheless, global, direct, and his interpretations appear to him as self-evident, although the operational structure had become the essential texture of these reactions. A certain answer may be based on operational schemas and, despite this, display the properties of an intuitive cognition; it appears subjectively non-explicitly justified and *a priori* evident.

As has already been mentioned, in Piaget's terminology, the word "intuitive" means everything which is not formal. Consequently, Piaget does not see any contradiction between considering a cognition intuitive and at, the same time, classifying it as operational. With that attitude we fully agree. But

Piaget does not explicitly distinguish, at the concrete-operational level, cognitions which are intuitive and cognitions which are operational without being intuitive.

An eight-year-old child is asked to perform the addition 8 + 5. He does not possess an intuitive, ready-made answer. He has to calculate (for instance by counting 8 + 1 + 1 + 1 + 1 + 1 or by calculating 8 + 2 = 10; 10 + 3 = 13 etc.). This is an analytical solution not an intuitive, global one based on self-evidence — although the child understands intuitively the operation of addition. When asked instead to solve 2 + 2, the child will probably reply instantaneously: 4.

This answer expresses a learned relation but also, probably, the intuitive, global representation of two joined sets $\{\cdot\cdot\} \cup \{\cdot\cdot\}$.

We consider that the distinction between an answer which is operationally based and at the same time intuitively accepted and expressed, and an answer which is based on operational schemas and is also analytically, explicitly justified is important both psychologically and didactically. An intuitively accepted solution, as is frequently emphasized here, represents a more direct, and much deeper involvement of the individual than an analytical solution without an intuitive basis. It is didactically important to detect the intuitive basis of a student's attitude with regard to a problem. If the intuitive representation is scientifically adequate one may fruitfully build further conceptual structures on it. If the intuitive basis is not adequate, conflict situations may be generated.

Let me give another example taken from the Piagetian studies. The question refers to the equilibrium in the balance.

Rol (10; 10): "*You have to change the position of the sack because at the end it makes more weight* . . ." (Inhelder and Piaget, 1958, p. 171). The child understands intuitively, globally the relationship between weight and the distance to the fulcrum.

Fis (10; 7) sees that P does not balance F "*because it is heavy: that one* $\{F\}$ *is too light*" — What should be done? "*Move it forward*". He moves P toward the axis and attains equilibrium (*ibid.*, p. 171).

The child understands *intuitively that, at the same side of the balance*, weight and distance are inversely proportional. These children's solutions imply the composability and reversibility of operations, characteristic of operational reasoning. But there is no quantitative, explicit formal analysis of the situation by the child. All of these are examples of operational intuitive reactions.

Intuitions do not disappear at the formal operational stage. A formally presented and justified solution may imply or may not imply an intuitive acceptance of the solution concerned. If an adolescent or an adult is asked about the price of 3 liters of juice if one liter costs 0.9 dollars he knows that a multiplication has to be performed. He feels it intuitively and he is able to justify and perform it numerically. If the question is about 0.90 liter of wine and 1 liter costs 3 dollars, the problem seems more complicated, and most students fail at this type of question (in which the operator is a decimal). Some do not answer at all, some students suggest division. Even if one knows that multiplication is the correct answer, the solution usually lacks an intuitive, sympathetic acceptance (see Fischbein *et al.*, 1985). It has to be remarked that in both problems the arithmetical solution consists of multiplying 0.90 by 3. The difference is in the role of the decimal. In the second problem the decimal has the function of the operator. The intuitive model of multiplication remains repeated addition and consequently a multiplication in which a decimal is the operator has no intuitive meaning.

According to Piaget:

> There is . . . a group of natural capacities also called "intuitive", the initial stages of which can also be considered as corresponding to lived-through experiences, and whose later stages are in their turn characterized by operations more and more "abstracted" from material action (by "reflective abstraction"), but which are more and more independent of any form of representation: they are the operational intuitions concerned with discrete objects. We may cite as examples the "intuition of $n + 1$" brought in by Poincaré to justify the so-called primitive character of numerical iteration; Brouwer's intuition of many-one, which presupposes an operation of colligation; the intuition of the transfinite in Denjoy's sense (the convergence to a limit in the series $1 + \frac{1}{2} + \frac{1}{4} + \ldots$) and, in a general way, all that we call intuitive in the elementary handling of classes, relations and numbers concerned with discontinous elements (Beth and Piaget, 1966, pp. 219—220).

The above examples, except the last, are primary intuitions in our acceptation. Although some knowledge is required, these intuitions have natural roots. When the relevant knowledge is acquired (in this case, elementary arithmetical notions) the students consider it natural and self-evident to give the expected answers. Kant affirms that such solutions are based on synthetic *a priori* judgements. This is the reason why they appear self-evident. The only exception in Piaget's list is represented by the last example. According to our findings, people do not accept naturally that $1 + \frac{1}{2} + \frac{1}{4}$ equals 2. There are two different intuitive answers. One is that the "sum is smaller than 2"; the second is that "it tends to infinity" (Fischbein, Tirosh and Melamed, 1981, p. 499).

It has been found that subjects aged 12 and more (formal operational period) possess a correct, natural, intuitive understanding of the following probabilistic concepts: the concept of chance and of the quantification of chances as the relationship between the number of favorable and of all possible equally likely outcomes; the fact that increasing the number of conditions imposed on an expected event diminishes its chances (which corresponds to the multiplication of probabilities). By contrast, there is no natural understanding of the compound character of some categories of events nor of the necessity to inventory the different situations which can produce the same event (for instance, when throwing a pair of dice, there is no intuitive understanding of the fact that there is a difference between the probabilities of getting the pair 5-5 and the pair 5-6) (Fischbein, 1975, pp. 138–155). (Even eminent mathematicians can make this sort of error. According to Pedoe (1958), D'Alembert believed that if a coin was tossed twice, than each of the three outcomes — heads 0, 1 or 2 times — was equally likely.)

The category of *secondary intuitions* implies the assumption that new intuitions, *with no natural roots*, may be developed. Such intuitions are not produced by the natural, normal experience of an individual. Moreover, very often they contradict the natural attitude towards the same question. According to our primary intuitions, we tend to consider that in order to keep the velocity of a moving body constant, a force is necessary. According to that representation, when a body is launched it acquires a certain impetus, an invested force. It keeps moving until the impetus dies out. According to Newtonian physics, on the contrary, a body preserves its state of rest or rectilinear constant motion if no force intervenes. This is the principle of inertia.

Intuitively, it is difficult to accept such an interpretation. If that interpretation can be transformed from a learned conception into a belief then we refer to it as a secondary intuition. Such a belief will never be acquired naturally in the normal conditions of our terrestrial life.

If for a mathematician the equivalence between an infinite set and a proper sub-set of it becomes a belief — a self-explanatory conception — then a new, secondary intuition has appeared.

The statement: "the sum of the angles of a triangle is 180°" is not self-evident. One accepts it by proof. First of all, it is surprising to learn that, no matter what shape the triangle has, the sum of its angles remains constant. If, by certain means, we become able to see *directly* that the sum *must necessarily* remain constant (because of inner compensations) we have acquired a new intuitive understanding — a secondary intuition. (Such a means might be provided by Papert's Turtle Geometry based on LOGO.)

William Feller writes in his book *An Introduction to Probability*:

> In contrast to chess, the axioms of geometry and mechanics refer to an existing intuitive background. In fact, geometrical intuition is so strong that it is prone to run ahead of logical reasoning. The extent to which logic, intuition, and physical experience are interdependent is a problem into which we need not enter. Certainly intuition can be trained and developed. The bewildered novice in chess moves cautiously, recalling individual rules, whereas the experienced player absorbs a complicated situation at a glance and is unable to account rationally for his intuition. In like manner mathematical intuition grows with experience, and it is possible to develop a natural feeling for concepts such as four-dimensional space (Feller, 1957, p. 2).

These intuitions to which Feller refers in the last lines represent, in our terminology, *secondary intuitions*. The fact that a mathematician accounts for their existence as *intuitions* not as mere formal acquisitions is an argument which supports their real existence.

Felix Klein (1898) has used the term "refined intuition" and F. Severy has written about "second degree intuitions" (1951). Patrick Suppes writes about the importance of developing intuitions for finding and giving mathematical proofs:

Put in another way, what I am saying is that I consider it just as necessary to train the intuition for finding and writing mathematical proofs as to teach intuitive knowledge of geometry or of real number system (Suppes, 1966, p. 70).

Hans Hahn, who has sharply criticized the use of intuition in mathematics, writes:

If the use of multi-dimensional and non-Euclidean geometries for the ordering of our experience continues to prove itself so that we become more and more accustomed to dealing with these logical constructs; if they penetrate into the curriculum of the schools; if we, so to speak, learn them at our mother's knee, as we now learn three-dimensional Euclidean geometry, then nobody will think of saying that these geometries are contrary to intuition. They will be considered as deserving of intuitive status as three-dimensional Euclidean geometry is today (Hahn, 1956, p. 1976).

Hahn implicitly admits in the quoted lines that intuition can change, that higher-order intuitions can be formed by adequate instruction and, finally, that correct, scientifically validated intuitions are an essential complement to the conceptual framework in science and mathematics education.

It is also important to note that Suppes does not write on "teaching the axiomatic method" but rather on "training the intuition for finding and writing mathematical proofs, etc." He thus implicitly accepts that, in order to build an axiomatic theory (i.e., a formal, non-intuitive and sometimes anti-intuitive one), *what we firstly need are adequate, efficient intuitions.*

Piaget uses the term "pure intuition" for cognitions which are completely detached from any practical experience and which are the product of higher order forms of reasoning. He mentions the French mathematician Bouligand who uses the term "intuition prolongée" in order to describe the way in which one builds the passage from three to four or *n* dimensions by analogy with the passage from two to three dimensions and by generalizing from a double to a triple integrale (Beth and Piaget, 1966, p. 223). Piaget mentions also the Cantorian construction of the transfinite sets in the same sense.

These too are secondary intuitions in our terminology. These examples seem to indicate that secondary intuitions may present various degrees of abstraction, sophistication and complexity.

Our term "primary intuitions" does not imply that these intuitions are innate, or *a priori*. Intuitions, both primary and secondary, are in fact learned cognitive capacities in the sense that they are always the product of an ample and lasting practice in some field of activity.

The distinction between primary and secondary intuitions is not an absolute one. One may rather consider a continuum ranging from very elementary, naturally acquired, intuitive cognitions (for instance the intuition of the permanence of an object, or the conservation of quantities) to very complex, genuinely counter-intuitive notions (like *n*-dimensional space or the relativistic interpretation of simultaneity). Between these extremes there is an infinity of nuances of cognitions acquired more or less naturally or more or less against our natural intuitive tendencies. It is, for instance, easier today to get used to the Newtonian understanding of inertia — which was originally counter-intuitive — than to the relativistic interpretation of space and time.

The relativity of the distinction between primary and secondary intuitions refers also to the fact that it depends on the cultural environment of the individual.

Our primary space intuitions, for instance, are different from those of people belonging to a different culture. In this respect, Alan Bishop quotes some striking examples in his paper devoted to 'Visualization and Mathematics in a Pre-technological Culture' (1978). For instance, he writes that for the Paiela (a Papua-New Guinea highland group) "space is not a container whose contents are objects. It is a necessary dimension of the objects themselves." For another group, the Kamano-Kafe of the Eastern Highlands, the four units of length are "long," "like long," "like short," and "short." Bishop concludes that "our conceptions of space with its items of objective measurement are not universal nor are they 'natural', 'obvious', or 'intuitive'. They are shaped by our culture. They are taught, they are learnt" (Bishop, 1979). We fully agree with Bishop's affirmation that spatial conceptions are shaped by the cultural environment — that they are taught, they are learnt. But we disagree with the first part of his statement (that space representations are not "obvious", are not "intuitive"). In fact, in the context of a certain culture, these "space conceptions" *do* appear as being "natural", "obvious", and "intuitive". In the educational process, we have to take into account the "obviousness" of these intuitions. We cannot say that we may neglect the existence of such intuitions because they are not natural *in an absolute manner*. When I buy 500 g cheese, the weight of the cheese is for me a reality, despite the fact that weight is only relative. The same cheese has no weight in an artificial satellite. From an epistemological standpoint, it is, of course, a fact of fundamental importance that intuitions are not *a priori*, genetically built-in truths. *But from a psychological, educational point of view, it is also of fundamental importance to identify categories of interpretations (correct or incorrect) which appear self-evident and imperative* (despite the fact that they are so only in the realm of a certain culture).

SUMMARY

Two classifications have been proposed to clarify the complex domain of intuitive cognitions. The first is based on the roles played by intuitions of different types in relation to other cognitive activity. Four types were distinguished on the basis of this criterion, as follows:

1. *Affirmatory intuitions*, which can be further categorized into (a) semantic, relational, inferential, and (b) ground and individual intuitions.
2. *Conjectural intuitions*, with respect to which a further distinction can be made between novices and experts within specific domains.
3. *Anticipatory intuitions*, which are implicated in problem-solving, as are ....
4. *Conclusive intuitions* (See also schematic diagram on p. 64).

The second basis of classification relates to the origins of intuitions. *Primary intuitions* are those which develop on the basis of normal everyday experience (which is of course subject to cultural variation). *Secondary intuitions*, by contrast, are those which are acquired, not through natural experience, but through some educational intervention. Often these are inconsistent with the corresponding primary intuitions relating to the same concepts.

CHAPTER 6

INFERENTIAL INTUITIONS AND LOGICAL REASONING

In this chapter, one of the categories of intuitions identified in the previous chapter, namely inferential intuitions, is discussed in detail. Thom (1971) provides an example of an inferential intuition:

> The fox knows that if the hens are in the henhouse and the hen-house is in the yard, then the hens are in the yard; he does not bother with set theory. Everyone uses set theory from the moment he exists, just as M. Jourdain in Molière's Le Bourgeois Gentilhomme uses prose without knowing it (p. 699).

Inferential (or logical) intuitions are those which express the feeling of validity which accompanies logical operations. They are intrinsically involved in reasoning and can be either primary or secondary intuitions. Inferential intuitions play a fundamental role in scientific and mathematical thinking and, consequently, in science and mathematics education.

In a syllogism, the conclusion is determined by the premises. *But the validity of the syllogism as a method of deducing a truth from previously accepted premises cannot be proved.* (Lewis Carroll (1895) see also Hofstadter, 1979) showed how an infinite regress is created by trying to logically justify the modus ponens rule according to the same logic as operates within that rule.) We must accept it by intuition (Ewing, 1941, *cf.* Westcott, 1968, pp. 17—19). This is an example of inferential intuition.

It is by intuition that we accept the universality of inductive inferences. If, in a number of trials, it has been found that iron is electrically conductive, we tend to generalize this finding, and conclude that iron is *in general* electrically conductive. What is the basis for such a generalization? No explicit proof can be found for the validity of this operation. The only thing which can be said is that, frequently, a generalization from a finite number of findings to a universal statement has not been contradicted by facts (but this argument itself appeals to induction).

Everything that has been said about logical inferences with reference to empirical facts is also valid for mathematical reasoning. For instance, what has been called "mathematical induction" is based on an intuitively accepted conviction that, on some mathematical grounds, extrapolation is legitimate.

There is some information available concerning the development of the intuitive understanding of logical operations. It has been shown that concrete operational children are able to identify the conclusion which follows from given premises in a categorical syllogism (of the form AAA or EAE). It is more difficult for them to formulate that conclusion by themselves. For instance, on the table in front of the child, there are four red squares, a yellow triangle, a blue triangle, a yellow rectangle and a blue circle (all of them of plastic material). The experimeter asks: "What is the shape of the red figures?" The normal answer is, of course: "All the red figures are squares" (the first premise). The experimenter hides the figures from the subject's eyes, removes the non-red figures, and covers the remaining ones (red) with a grating. Through the grating, the shapes of the figures are no longer visible, but the color can be identified. The experimenter asks: "What color are the figures underneath?" The normal answer is: "All the figures under the grating are red." This is the second premise of the syllogism. Then the experimenter asks: "What is the shape of the figures under the grating?" The subject is asked to answer in two different ways: (a) to verbally formulate the conclusion, and (b) to choose, from four written sentences, the correct one (i.e., the figures under the grating are squares). This example embodies the mood AAA of the first syllogistic figure. In the same manner, all four moods of the first and of the second syllogistic figure were investigated. The subjects were pupils from grades 2, 3, 4, 6, and 8 (20 pupils for each age) (Fischbein, Barbat and Minzat, 1975).

It has been found that 80% of the children of ages 7—8 (grade 2) were able to identify the correct conclusions in an AAA type. In syllogisms of the form EAE and AII, there were 65% correct answers, while on the mood EIO, for both the first and the second figures, there was only one (5%) correct answer (which may well have been by chance). It may be concluded from these data, *that there is a natural, logical competence concerning categorical syllogisms of the form AAA, in concrete operational children.* The child concludes that under the grating there are squares, even though he does not see them. This kind of belief is representative of an inferential intuition.

Our point is that a syllogism, as well as any other logical inference, is not a pure conceptual structure. *It always expresses a more basic extra-logical attitude which is the belief in the validity of that inference.*

Consider this piece of hypothetical reasoning: (1) If object A is a metal then it will conduct electricity; and (2) Object A is a metal (it has been identified for instance, as being sodium); then, (3) Object A conducts electricity. *Nothing here is intuitively evident except the validity of the inference.* It may be argued that this represents just a mental habit. As a matter of fact it is not only a mental habit; *it is a type of cognition accompanied by a feeling of intrinsic conviction.* The algorithms for multiplying or dividing two numbers are also learned and they finally become mental habits — but they do not have the intrinsic evidence of an intuition.

Generally speaking, we may safely say that the axioms of logical thinking are, in fact, based on such fundamental beliefs. They constitute the domain of logical intuitions. Mathematical education (and intellectual education in general) should not be satisfied with training blind automatic intellectual skills corresponding to the formal laws of logical thinking. Such blind rules do not work by themselves in an actual problem-solving process. They may work in solving blind exercises. We can teach a pupil the truth table of implication; it does not follow that he will use it naturally in a thinking process if the corresponding intuitions have not been developed. Let us take an example: (1) If quadrilateral A is a square then its diagonals are equal, and (2) A is *not* a square. Pupils naturally tend to conclude that A does not have equal diagonals. On the other hand, from the statement: "It has been proved that the diagonals in figure A are equal," pupils tend to conclude that A is a square. When using an implication $p \rightarrow q$, children aged 12—13 do not distinguish naturally between the *uncertain* conclusion which can be drawn by affirming q and the *certain* rejection of p which follows from the negation of q.

Again, the educational problem is not only that of building a set of mental skills for logical thinking. New intellectual beliefs, i.e. intuitions, have to be built. Their role is not only to suggest or to confirm inferences. Their function is also to measure, logically as well as subjectively, the completeness, or conversely, the incompleteness of an argument: "I feel that something is wrong with my reasoning". Such feelings are possible only if sound active inferential intuitions have been built.

A relatively rich bibliography concerning the development of logical structures is available, including the work of Piaget. We have already mentioned the concept of operational intuitions developed by Piaget. The difficulty is that Piaget, as we have already mentioned, does not systematically identify the specific intuitive facets, the intuitive constraints and limitations related to logical reasoning in operational children.

With regard to syllogistic reasoning in children the findings are not unequivocal. Hill (1961) has claimed that concrete-operational children may exhibit syllogistic reasoning. Others like Roberge and Paulus (1971) consider that, in general, children are not able to use syllogistic reasoning. Most of the authors (Ennis *et al.*, 1969; O'Brien and Shapiro, 1968; Peel, 1967; Taplin *et al.*, 1974) suggest that children may sometimes reach correct conclusions in syllogistic reasoning but they are not consistently correct. Knifong (1974), referring specifically to conditional reasoning, claims that, in fact, children answer correctly only if the correct solution may be found by *transduction* and this may happen with the forms of reasoning called *modus ponens* and *modus tollens.*

> For instance: "If this object is sugar then it is sweet".
> *Modus ponens*: "This object is sugar — then it is sweet".
> *Modus tollens*: "This object is not sweet, then it is not sugar".

Knifong claims that a child reaching the correct conclusion in the above cases does not exhibit implicit but rather transductive reasoning. The elements are connected in the child's mind as a non-directional juxtaposition (through a kind of bilateral relationship). In order to prove this assertion one has to check the other two forms of conditional reasoning: the negation of the antecedent (this object is not sugar), and the affirmation of the consequent (this object is sweet). If the child really understands the relation of implication, he has to understand that in these two instances no clear-cut conclusion may be drawn. If the child's reasoning is transductive he would draw a negative conclusion when the antecedent is denied and an affirmative conclusion when the consequent is affirmed. It seems that the findings cited by Knifong are consistent with these predictions.

As a matter of fact, things are even more complex. One may consider at least three types of experiments: (a) The questions are put in a symbolic form: "If *p* then *q*; *p* happens; what about *q*?"; (b) The questions refer to concrete, familiar data (if this is sugar then this is sweet etc.); (c) The questions refer to concrete but purely conventional relationships (in my garden if the bug is big, then it is striped).

As we have said, our hypothesis is that the validity of various types of logical inferences may be accepted intuitively by direct apprehension.

Trying to check such an hypothesis, the researcher faces a great difficulty. Let us consider that the question is put in a purely symbolic form. (For instance: "if *p* then *q*"; we know that *p* is true; what is the conclusion with respect to *q*?) It is possible that the subject will not be able to draw a correct conclusion simply because the abstract form of the propositions may prevent the logical schemas being elicited — not because the subject does not possess these schemas.

On the other hand, if one confers concrete meanings on the symbols the subject may draw his conclusions not through pure logical means, but by directly considering the concrete situation. For instance: "If this liquid is water you are allowed to drink it. But this liquid is not water. What is the conclusion?" A subject may answer correctly that no clear conclusion may be drawn, either because he knows that in an implication $p \rightarrow q$, by denying *p* one *cannot* draw a certain conclusion about *q*; or because he simply knows that if the liquid is not water. it may nevertheless be some other drinkable liquid.

This methodological difficulty has led to apparently contradictory conclusions concerning children's ability to correctly use various logical inferences. In order to overcome this difficulty, Kuhn (1977) used three types of experiments in relation to conditional reasoning. In the first study, the syllogisms referred to familiar notions. (For instance: "All the boys in Tundor play marbles. Chrys does not live in Tundor. Does he play marbles?")

In the second study, concrete aspects were used again but the relations between them were completely conventional; no specific life experience could suggest to the pupils the correct answer.

In the third study, the subjects were presented with the following story. In a garden there were three kinds of bugs: a big striped one, a small striped one and a small black one.

The subjects were given 8 sentences and they had to decide which of the sentences was correct according to the previous description of the existing bugs. As a matter of fact, the above description expresses the implication: "big implies striped". The true sentences were: "In my garden, if a bug is big it is striped" and "In my garden if a bug is black it is small". The subjects had to identify these two correct sentences out of the eight presented. (An example of a false sentence is: "If a bug is big, it is black)

By comparing the results obtained in the above studies we may hope to get an insight of the spontaneous capacity of children to use logical schemas.

Kuhn's general conclusion was that the logical operations required for solving traditional syllogisms are acquired by middle childhood. The poor performances among children and adults on traditional syllogism tests are due to the structure of the tests, and not to a lack of competence in logical operations (Kuhn, 1977).

Let me cite one more example from Kuhn's paper. Several subjects, in deciding whether "Dave, who has blond hair, lives in Tundor — given that all people in Tundor have blond hair", first responded "yes" and then immediately corrected themselves, saying, for example: "No, wait a minute, I have blond hair and I don't live in Tundor". Kuhn concludes: "The real life meaning of these propositions was evidently necessary in this case to enable a child to recognize the possibility of $\overline{p}$ in conjunction with q" (Kuhn, 1977, pp. 348–9).

An essential methodological and theoretical problem is raised by Kuhn's comment. Are we entitled to claim that the child in the above example finally concluded correctly because he implicitly knows the truth table of implication? In fact it seems likely that his real source is extra-logical (he has blond hair himself although he does not live in Tundor). The relatively high percentage of correct answers (even for second and third graders) for the critical (non-concluding) items (denying the antecedent, affirming the consequent) may be due to the extra-logical sources of information on which the child can rely. Bereiter *et al.* (1979) also stressed the fact that one has to distinguish, when evaluating a child's capacity to perform logical reasoning, between strictly logical inferences and inferences made by using extra-logical data.

Kuhn's other two studies mentioned above support the same conclusion. She found that with tests which do not allow the subject to rely on some practical information (lying beyond the boundaries of the logical constraints) children's performance may be very poor on some critical items. For instance, for $p \rightarrow q$ items in study 2, Kuhn found no correct answers for third and fifth grader's and 10% correct answers for seventh graders. Similar conclusions have been drawn by O'Brien *et al.* (1971), Bereiter *et al.* (1979) and Adi *et al.* (1980).

Briefly speaking, one may conclude that some stable, basic logical structures develop in concrete operational children. The first to develop are those expressed in categorical syllogisms of mood AAA and in conditional syllogisms of the *modus ponens* type — both related to class inclusion. Other types of inference develop later. But it seems that certain types of inference (such as the $\bar{p} \rightarrow q$ form) are never completely assimilated.

There are still other problems with important theoretical and practical implications.

What are the relationships between logical schemas and the corresponding practical competence? Is the fact of knowing the logical rules, required for solving a certain problem, a sufficient condition for solving that problem? The Piagetian conception seems to imply a positive answer to this question, with the following essential restriction: one has to be able, first of all, as a natural effect of age, to use the logical operations in a combinatorial way, as, in fact, happens at the formal operational stage (Inhelder and Piaget, 1958).

There has been considerable criticism of the Piagetian view (e.g. Johnson-Laird, 1983; Parsons, 1960; Wason, 1977). Johnson-Laird (p. 24), in his criticism of what he terms "the doctrine of mental logic" quotes as the extreme form of this doctrine the statement of Inhelder and Piaget (1958, p. 305) that "reasoning is nothing more than the propositional calculus itself". Later he cites evidence (as does Wason (1977)) to support his contention that "subjects do not spontaneously examine the combinatorial possibilities in a systematic and exhaustive fashion" (p. 46).

The point of view expressed in this book is different from that of Piaget. Genuine practical competence depends on various factors among which two are directly relevant to the present study. One is the influence of practice. The second is represented by the intuitive constraints acting in the respective circumstances.

The *fact that one knows the truth table of implication does not determine, by itself, the correct use of it in every situation.*

The fact that one knows *formally* a certain logical rule does not imply that it has been intuitively assimilated. The fact of knowing de Morgan's laws:

$$\overline{p \cdot q} = \bar{p} \vee \bar{q}$$

$$\overline{p \vee q} = \bar{p} \cdot \bar{q}$$

does not mean that one regards these laws as evident. In contrast, given $A = B$ and $B = C$ one sees directly, as an evident and necessary inference, that $A = C$.

An interesting experiment conducted by Piaget and Bullinger (Piaget, 1980) has shown that, beginning at a certain age, transitivity is grasped as an evident property of equivalence.

Children, aged from 5 to 12, were asked to compare a series of 7 disks the diameters of which increased progressively by steps of 0.2 mm. This is a practically indiscernible difference. The subjects were allowed to compare the disks only by pairs. They concluded, based on transitivity, that the first and the last disks should be equal. The subjects were surprised to find that the two extreme disks were in fact different in diameter. Only eleven-year-old children could solve the conflict, suggesting that all the disks of the series were, in fact, unequal and that they were ordered in series. What we want to emphasize is the fact that the children were *surprised* by the inequality of the extreme disks. It is this kind of surprise when a certain representation is contradicted which may be considered as an indicator of the intuitive nature of that representation.

Let me mention another example. One is shocked when one learns that the set of the natural numbers and the set of even numbers are equivalent. A set cannot be logically equivalent with one of its proper subsets. It is an intuitive, logical rule that the whole is bigger than each of its parts. When a statement seems to contradict this rule one feels that evidence itself is contradicted.

As we have already said there is no systematic experimental data available concerning the intuitiveness of various logical rules. It seems that researchers have not been concerned so far with this question.

The intuitiveness of logical schemas might be considered unimportant if one used logical procedures only for checking solutions already attained.

But logical inference has to play a fundamental role in every intellectual, creative endeavor, very often implicitly. This means that the intrinsic evidence — or the lack of evidence — of the properties of a certain logical operation may have a direct — uncontrolled — impact on the very course of a reasoning process.

Let us consider an example, the famous four card problem: Since its introduction, this problem has given rise to a very substantial literature (see, for example, Evans (1982); Johnson-Laird (1983, p. 30ff). In one form of the problem, the following rule is stated: "If a card has a vowel on one side, then it has an even number on the other side". Cards are presented showing E, 4, K and 7. The subjects are asked to respond by "yes" or "no" whether they need to know what is on the other side of each card in order to find out if the rule is true or false. In a replication of the experiment performed by Wason and Evans (1975), it was found that out of 24 college students questioned, none was able to choose the correct cards. Adi *et al.* (1980), in a more recent version of the experiment, found that 14.9% college students answered correctly with regard to all four cards presented (that is, one has to turn E and 7; 4 and K are not relevant).

In the above example one deals with an implication: p (a card has a vowel on one side) implies q (the card has an even number on the other side). No extralogical considerations may help. Formally $p \rightarrow q$ means that affirming p one implicitly affirms q and that denying q one implicitly denies p. Denying p or affirming q do not lead to certain conclusions. Consequently — returning to the four cards — there are two relevant checks to be performed: card E (corresponding to the affirmation of p) and card 7 (odd number = the negation of q). Generally, the full correct solution is not found even by people who know the truth table of implication. Our hypothetical explanation is that these people have not assimilated *intuitively* the complete structure of implication. To assimilate intuitively means, according to our conception, to get the respective concept turned into an intrinsically obvious and behaviorally meaningful, efficient cognition. One may, plausibly, assume that only the *modus ponens* component of implication naturally acquires an intuitive interpretation and this would mean that implication is converted intuitively into a categorical AAA syllogism (with the conditional structure eliminated or converted into a relation of equivalence).

For instance, some of the subjects in the Adi *et al.* experiment answered in the following way: "I don't need to turn card E, because there will be an even number on the other side", or: "No need to turn card 4 because there will be a vowel on the other side". There were students who answered: "I want to turn card 4 to see if there is a vowel on the other side" (a check which would be relevant for equivalence but not for implication).

The formal structures of reasoning supposed to represent a complete combinatorial system in adolescents (Inhelder and Piaget, 1958) are only potentialities considered from the point of view of practical competence. Some of them simply have to be learned. They do not develop spontaneously at the formal operational stage. Some of them, even after being learned and known formally, do not spontaneously become an integral part of the current mental behavior; they may be known conceptually, without being really influential as tacit, implicitly productive mental tools. They may even conflict with primitive logical structures.

The methodological difficulty of getting accurate data about the degree of intuitiveness of a logical schema is certainly increased by the fact that these schemas may reveal the presence or absence of their intuitiveness only when referring to some content. It is also plausible that this degree of intuitiveness is not absolute (i.e. absolutely related to the nature of the logical schema itself); it may also depend on the content (even if the subject may not have the possibility to refer to his personal experience in order to get the correct answer). Strong but not entirely consistent effects relating to the content have been demonstrated in a number of studies using variants of the 4-card problem (see Griggs (1983) and Wason (1983) for summaries of such studies). But is it possible to produce items referring to a certain content and at the same time to make sure that this content is conventional to such a degree that it cannot suggest any extra-logical consequences?

All of these are open questions. We do not possess systematic experimental evidence. But the problem of the intuitiveness of the various logical connectives is of fundamental importance for mathematics and science education. As we have several times emphasized, the fact of knowing the truth table of a logical operation does not ensure its automatic use in practical problem solving situations. This affirmation is highly relevant, especially for mathematical thinking, where very often the decisions about the validity of certain inferences have to be taken *without the possibility of using extra-logical considerations.* Such verifications may be performed sometimes by empirically checking the consequences of certain hypothetical statements after one has reached a conclusion. But very often, during the solving endeavor itself, one simply does not have the possibility (or even the idea) to perform such checks.

Let us consider a few more examples. In research carried out by Galbraith (1981), pupils were asked about numbers for which the sum of the digits can be divided by 7. One provides examples: 34 $(3 + 4 = 7)$; 185 $(1 + 8 + 5 = 14)$. The questions continue like this: "If we make a list L of all such numbers which are less than 70, the start of it looks like this: 7, 16, 25, 34. Write down the next largest number in the list. Gary says: If you start with 7 and keep adding nines you always get a number in the List L.

(1) Is Gary right?

Brenda says "Every number in the list can be found by adding 9 to the previous number. You start with 7".

(2) Is Brenda right? (Galbraith, 1981, pp. 9—10).

In order to get the right answer with regard to the statements of Gary and Brenda, one has either to find corresponding proofs or to check empirically every number in the list. As a matter of fact, many pupils (12—15 years of age) were not able to accurately use the logical structures needed for solving the problem.

Galbraith cites answers like: "If a rule goes for one, it will go for another"; or: "If it works for three it should work"; "Brenda is right to a certain extent"; "One example is not enough to disprove it"; "Could be a freak accident in a million chance". These pupils simply do not use implication, as a full logical tool. In order to affirm, with Brenda, that all the numbers in the list L (i.e. the numbers divisible by seven) may be obtained by starting from 7 and adding successively 9, one has, in the absence of a formal proof, to check for counter-examples, that is by invoking the *modus tollens* component of implication. And this many subjects have not done spontaneously. But even after having found numbers in the list L (59 and 68) with the sum of the digits divisable by 7 but which could not be obtained by successively adding 9, many subjects did not accept that the statement of Brenda is thereby negated (that is $\bar{q} \rightarrow \bar{p}$). The students' approach is, rather, an empirical one, according to which exceptions may not refute a law (because, potentially, some uncontrolled factors may have interfered).

O'Brien *et al.* (1971) have found that only 20% of grade 10 students were able to answer some implication tests correctly. They concluded that this may explain the lack of success of many students in constructing a mathematical proof or checking its validity.

According to Piagetian theory, the main characteristic of the formal operational stage is the presence of a combinatorial system of the basic logical operations. In other words, if an adolescent finds, for instance, that a property p and a property q appear together he will normally and spontaneously ask whether one has to do with an implication ($pq \vee \bar{p}q \vee \bar{p}\bar{q}$), with an equivalence ($pq \vee \bar{p}\bar{q}$), with a disjunction ($pq \vee \bar{p}q \vee p\bar{q}$), with a conjunction, etc. Certainly he will not refer, explicitly, to the logical terminology but, in Piaget's conceptions, he will strive to produce the experimental conditions which will enable him to make the right choice. (As has already been mentioned, this view has come in for a great deal of criticism.)

Most of the existing evidence supports a different conclusion, at least with regard to mathematical thinking in adolescents. It seems that, without a direct intervention of empirical information, many subjects are not able to spontaneously rely only on their logical schemas for drawing correct formal conclusions. This affirmation holds especially for conditional reasoning.

Our explanation is that, in many cases, the respective logical schemas are ineffective not because the subject ignores them but because they have not been assimilated intuitively.

The whole problem needs much more investigation but one conclusion seems to be clear. The training of logical capacities is a basic condition for success in mathematics and science education. We refer not only to a formal-algorithmic training. The main concern has to be the conversion of these mental schemas into intuitive efficient tools, that is to say in mechanisms organically incorporated in the mental-behavioral abilities of the individual.

PART II

FACTORS WHICH SHAPE INTUITIONS

CHAPTER 7

INTUITION AND EXPERIENCE

THE BEHAVIORAL ROOTS OF INTUITIVE REPRESENTATIONS

Experience is a fundamental factor in shaping intuitions. There is little systematic evidence available supporting that view, i.e. evidence demonstrating that new intuitions can be shaped by practice. However, there are introspective and other general empirical descriptions and also theoretical analyses supporting the view that the *basic source of intuitive cognitions is the experience accumulated by a person in relatively constant conditions.*

Three main experimental aspects relevant to intuition may be identified:

(a) The general, common elements of human experience.
(b) The aspects of experience related to the particular geographical and cultural environment in which a person lives.
(c) The particular practice of the individual related to various domains of his life (for instance professional intuitions).

To the first category belong the general space intuitions, which develop with age in every human being. As an effect of practice and, certainly, of biological maturation, the child learns to coordinate, in his first year of life, the various, initially heterogeneous, "spaces": the buccal, the tactilo-kinesthetic, the postural, the visual and the auditory. He thus gets a representation of space which is characterized by the permanence of objects and the coordination of displacements and positions in a unique frame (Piaget, 1967 and Beth and Piaget, 1966, pp. 213—214). On the basis of these sensory-motor acquisitions the development of the perceptual space capacities takes place; the child learns to compose and evaluate forms, dimensions, positions and distances. There are, certainly, some innate factors which contribute to the organization of spatial representations, but behavior and experience play a fundamental role. Piaget recalls Kohler's famous experiment. His subjects were asked to use glasses inverting the images of objects. After some days the subjects regained a normal image of the environment through adaptation.

Space subjectively gets the structure of a three-dimensional framework — the sense of anticipatory, behaviorally meaningful reactions. But this naturally shaped representation is not the Newtonian absolute space interpretation. Our natural space representations are non-homogeneous and anisotropic. One tends to dilate the zone under attention, while the periphery is contracted. One tends to amplify, perceptually, a furnished room as compared with an empty one, etc. One tends to attribute to space absolutely privileged directions — horizontal and vertical, up and down. One tends to represent space as "flat" (in the Euclidean—Newtonian sense) that is, with light spreading straight forward in all directions.

All these properties of naturally developed space representations are related to our terrestrial life and to our behavioral adaptive constraints.

At the same time spatial dimensions are naturally conceived as being absolutely reversible — in contrast to time. In principle, travelling from B to A is equivalent to travelling from A to B. If there are differences they are due to factors which are not attributable to pure space properties. This representation of space includes the tacit assumption of possible, objective, invariant properties — distances, forms, positions — and fixed points of reference. The basic mode of existence of external realities is subjectively that of solids — space is essentially a framework of solids, not of gases or liquids. Space and solids are somehow intuitively congruent because space is essentially represented as being stable, reversible, comparable, measurable. An intuitively acceptable and manageable system of reference is represented by rigid lines and surfaces, not by volatile entities.

Our intuitive representation of space is in fact a mixture of contradictory properties. It is a mixture of Euclidean—Newtonian properties and primitive beliefs. We refer, intuitively, to space as a milieu in which parallels may be drawn, in which measures may be performed (including the idea of objective, invariant distances), in which similarities and congruences may be established, in which notions like straight lines and minimal distances have a meaning. At the same time some directions appear, subjectively, tacitly to be non-reversible: "up" and "down", "horizontal" and "vertical". The space framework is subjectively, tacitly, *not* independent of some special directions and locations. Space is centered, with our permanent home, our town, our country representing the main subjective location references. Space seems to present an increasing density as one approaches the centration zones, with the effect that distances are increasingly amplified when approaching and, conversely, contracted when moving away from them.

What have all these to do with the concept of intuition? We are discussing space properties, and these belong essentially to the domain of perception. The reason for referring here to space is that it is *not* reducible to a conglomerate of sensorial images.

In fact, space representations and evaluations constitute a complex system of *conceptions* — although not necessarily formulated explicitly — which exceed the data at hand and the domain of perception in general.

Subjective space is an *interpretation* of reality not a *reproduction* of reality. It is shaped by experience but it exceeds experience. It is not only a product of experience but it is also a condition of experience, that is to say a condition for articulated, anticipated adapted reactions.

We have focused here on the intuition of space because it is a good paradigm for analysing various aspects of the relationships between experience and intuition in general.

There are, firstly, various sensori-motor schemas which develop as a part of the child's adaptive development. The child learns to coordinate his eye movements with his prehension reflexes and, thus, to obtain images of the surrounding reality which are behaviorally meaningful. He becomes able to structure the various categories of information — visual, haptic, kinaesthetic, postural — into images which are, at the same time, *representative* and *enactively* meaningful. These images process a three-fold character: they are stimulating devices, they are anticipatory devices and they are instrumental for feedback purposes.

The child sees a ball, he wants it, runs after it in accordance with the position of the ball and adapts his reactions to the motion of the ball. In a first account, one may say that the child perceives the ball and the moving image of it guides the child's movements.

In the above circumstances, there is much more in the child's representation of reality than a mere mirroring of a given reality. *It is a complex (of course, tacit) theory of space, as a framework and as an ensemble of properties, which lies behind the child's attitude.* When running after the ball, he anticipatively evaluates the trajectory of the ball. To run straight to the ball implies the notion of a certain privileged trajectory, the straight line, which corresponds to minimum distance, and likewise to minimum time and effort (to reach the ball).

The child not only *sees* the ball moving but he also *expects* that the ball will keep moving. He not only *sees* the ball but he also *expects* that it will continue to exist, to preserve its form, its magnitude and other properties. The child would be very surprised if the ball underwent a drastic transformation of form and magnitude. The child would be astonished if, wanting to come back with the ball to the initial point, he had to go or to run a much longer distance than was necessary for reaching the ball. (Obviously, we are referring to a child who, as an effect of age and experience, has already acquired these intuitions).

If the ball was first moving down a slope, the child would be extremely surprised if the ball suddenly started moving upwards by itself.

The child's representation of space is not a mere reflection of objectively given space properties. It is, rather, as I said, a complex, relatively coherent set of *expectations* structured in such a way that it may serve simultaneously as a guide for action and as a control system via feedback mechanisms. *It is this highly complex system of expectations, and programs of action, related to the movements of our body and its parts, which constitutes the intuition of space.*

As a matter of fact we are dealing here with *a cluster* of *intuitions*: distance evaluations and comparisons, expected trajectories, matrices of relative positions, all of these symbolizing, in fact, potential displacements with their potential effects.

Secondly, it is clear that the "intuition of space" is stored in an aggregate of skills, both sensori-motor and intellectual. As an effect of continuously accumulated experience, various images elicit certain adapted reactions. The adaptive meaning of these images is itself shaped by the personal practice of the child.

But skills are not intuitions. One may be very skilled in performing arithmetical operations, without displaying any special intuitions for that activity. A good marksman hits his target exactly; he is highly skilled. This is an effect of perfectly coordinated visual information and adapted reactions — but this is not by itself an intuition.

An intuition is more than a system of automatized reactions, more than a skill or a system of skills; it is a theory, it is a system of beliefs, of apparently autonomous expectations.

Experience has a fundamental role in shaping intuitions because, in certain circumstances, it shapes stable expectations.

Such expectations became so stable, so firmly attached to certain circumstances, that their empirical origin may, apparently, vanish from the subject's awareness.

This is, in fact, Hume's theory about the origin of the principle of causality. One expects that the same group of events will produce the same effects, because a "custom or habit" has been created as an effect of a repeated association.

In our opinion, there is much truth in Hume's interpretation. But, by adopting his view, one does not necessarily adopt his philosophical position as well. The idea that various categories of stable expectations have, psychologically, an empirical origin does not imply that these expectations do not correspond to objectively existing properties and relationships. This simply is a different problem. What we are claiming is that intuitions are based on stable, self-consistent expectations organized as beliefs, which are *apparently* autonomous in respect to particular empirical circumstances but are in fact, very often generated and shaped by long experience.

In other words, *experience may generate intuitions not only by generating stable patterns of reactions but also organized, apparently autonomous, systems of beliefs.*

When a child draws a straight line, he displays a skill. When he affirms that a line may be extended indefinitely, he expresses an intuition. This intuition is related to his experience. He has learned that he may extend the line as long as he wants. No absolute obstacle prevents him from continuing the operation. What he needs is enough paper, and sheets of paper may always be added (at least mentally).

A child has learned what multiplication means and that, by multiplying two numbers, the result obtained will be bigger than each of the terms. He has acquired a pattern of skills related to the operation of multiplication. At the same time, as an effect of his experience, *a belief* has been generated — the belief that *multiplication* (*necessarily*) *makes bigger.* This is no more a skill; this is a theory about multiplication.

The child who runs after the ball is able to perform this action on the basis of a multitude of acquired reflexes but beyond these well organized movements, there are the implicit beliefs that the ball cannot vanish, that one may minimize effort and time in reaching the ball by running straight, that there is no other way to get the ball than to cover the distance separating oneself from it.

All these are implicit convictions acquired through experience. They somehow represent the "theoretical background", the tacit, global, cognitive scene which makes possible, to a conscious human being, the coherent exercise of his patterns of skills in given circumstances.

EXPERIENCE AND INTUITIVE BIASES

The fact that intuitions are, to a large extent, shaped by experience explains why intuitions are very often instrumental in organizing and orienting our activity.

On the other hand, as has frequently been emphasized, intuitions may sometimes consist of distorted or erroneous representations of reality. This may seem paradoxical at a first glance if it is accepted that intuitions have a basically behavioral adaptive function.

In fact, the empirical origin and role of intuitions may partially explain the existence of intuitive biases *as well.*

The Terrestrial Limitations of Human Experience

First of all, human experience is necessarily limited in time, in space and in range of possibilities. Certainly, by intuition one *extrapolates* — and this is one of the fundamental virtues of intuitive cognitions - but the direction and the nature of extrapolations depend, nevertheless, on the conditions on which a person's experience is based, on the data it provides.

Our primary space representations are basically shaped by terrestrial life. We cannot imagine space either as being infinite or as absolutely limited. Both interpretations are totally outside the scope of our human experience. Consequently, our spatial intuitions — and our intuitions in general — are essentially finitist in nature. The notion of infinity (or *actual* infinity) leads to apparent logical contradictions.

We can hardly imagine a world in which "up" and "down" do not exist, a world in which things do not "fall down" but "stay where they are", even if they are totally unsupported.

By absolutizing the characteristics of "terrestrial space" one gets primary space intuitions; they are "correct" in the realm of terrestrial life.

One can hardly imagine that a body may continue to move indefinitely if no "impetus" intervenes to "push" it because such a "free" motion is impossible under terrestrial conditions.

It is difficult to intuitively grasp the idea that rest and uniform motion are only relative concepts because one may find a practically unlimited number of apparently fixed landmarks on Earth, i.e. objects rigidly attached to the ground! It is the ground itself which, first of all, appears to be the fundamental, the absolute image of immobility. To abandon this view — not only conceptually, but also intuitively — is an almost impossible task for an untrained mind.

The intuitive idea of the existence of absolute frames of reference was so deeply rooted in man's conception of space that it took two thousand years in the history of science until it could be replaced (with Galileo and Newton). But even *after* the Newtonian revolution, the primitive idea of the absoluteness of motion continued to influence the conceptions of the scientific community.

Newton himself, although he denied the existence of any absolute frame of reference in the universe, could not free himself from the fundamental intuitive need for such a framework. It was space itself to which Newton attributed the quality of absoluteness. To Newton, Absolute Space appeared as an infinite container which always remains "similar and unmoving". As a matter of fact, the idea of a totally empty framework, not marked by anything, is a mere conceptual fiction but it satisfies the fundamental need for an absolute (external, fixed) frame of reference shaped by our terrestrial life.

Still more interesting is the fact that the same intuitive bias continued to affect the scientific conception of reality hundreds of years after Galileo and Newton had established the relativity of states of rest and uniform motion. It was, in fact, this obstacle which prevented the scientific community from understanding the relativity of simultaneity which should have been a logical consequence of the principle of inertia.

Experience, then, plays a fundamental role in shaping our intuitions and this, as I have said, explains, at least partially, its impact on any productive, theoretical or practical endeavor. On the other hand, experience is always restricted to a limited system of circumstances and this contributes to limiting the domain of reliability and effectiveness of intuitions. *Nevertheless, intuitions, by their very nature and behavioral function, tend to appear, subjectively, as sure, self-consistent, universally valid representations.*

The Practicality of Intuitive Meanings

We have mentioned the necessarily *limited* nature of experience which imposes severe restrictions on the reliability of our intuitions.

A second aspect to be mentioned is that of *practicality*. By this term I understand the fact that one tends intuitively to attribute to notions and mental operations properties which, properly speaking, belong only to concrete, material realities.

Three properties relating to practicality will be discussed, namely (a) concreteness, (b) finiteness, and (c) the "duplication" obstacle. As an example of concreteness, it is impossible to imagine — intuitively — a particle otherwise than as a small marble and the motion of it as the motion of a marble. It is impossible to consider points or lines, even in a strict mathematical context, otherwise than through concrete embodiments. The fact that one describes a point as zero-dimensional or a line as uni-dimensional (in Euclidean geometry) does not completely eliminate the potential impact of such practical representations on mathematical reasoning. We have already mentioned various effects of this kind. Let us only recall the *intuitive* impossibility of accepting, for instance, that the sets of points of two segments of different length, or the sets of points of a segment and of a square (or a cube) are equivalent. *Intuitively*, it is impossible to accept that the two segments have the same number of points, because "intuitively" means dealing with "practical" points. As a matter of fact, the problem asked in this way has *no* intuitive answer, because "a point" is a concept, a construct, a mere definition with no real correspondent whatsoever. Cantor's contemporaries were perfectly aware of the abstract nature of mathematical entities. Nevertheless they were shocked when Cantor raised the problem of the possible equivalence between the set of points of a segment and that of a square! As a matter of fact, they had continued to think about points in terms of real, very small spots (although they would never have been willing to confess the substitution, not because of hypocrisy but because they simply were not aware of the substitution).

On the other hand, one is also tempted to consider that all infinite sets are equivalent (and, consequently, a relation of order among infinite sets has no intuitive meaning).

This example brings us to a second aspect of practicality (already mentioned previously). One is able, intuitively, to grasp objects, sets, or events which are either finite, or indefinitely extensible, but not *actually* infinite. That particularity of intuitive acceptance is certainly also due to the empirical origin of intuitive beliefs. Our logical schemas in fact represent coordinating devices of our experience — both practical and social — and therefore they must correspond to the general characteristics of the experience. One of these characteristics, clearly, is finiteness.

The objects we deal with practically are limited in space. The processes are limited in time, although, in fact, they may be imagined as indefinitely extensible. "Indefinitely extensible" is equivalent to what is called *dynamic infinity*. One may accept intuitively that a line can be extended indefinitely but this, in fact, implies that, at every moment, no matter how long after the process has started, what we get is, practically, a finite segment.

Before continuing, a remark is necessary. One generally distinguishes between intuitive forms of cognition and logical (analytical, deductive) inferences. It would, then, seem inappropriate to refer to the characteristics of logical schemas when discussing intuitive cognitions.

In fact, as we have frequently stressed, our logical schemas are subjectively based on intuitive grounds — which in turn are shaped by the constraints of our adaptive experience.

When dealing with actual infinity — namely with infinite sets — we are facing situations which may appear intuitively unacceptable. The classical example is the statement that if a set is infinite then it is equivalent with a proper subset of itself. For instance, the set of natural numbers and the set of even numbers are equivalent. The set of rational numbers and the set of integers are equivalent (i.e. contain the same "number" of elements). Such a proposition which affirms that a set and a part of it may contain the same number of elements (may be equivalent) is intuitively unacceptable.

We are intuitively not equipped to deal with *actually* given infinite sets. *Their* logic is not *our* logic, which is rooted in our practical experience.

A third aspect related to the practicality of our intuitions is the fact that it is difficult to perform mental operations which would be meaningless in practical terms.

An object cannot be practically, at the same time, in two different places. If John has 10 marbles and Richard has 7 marbles who has more? This comparison is practically and, of course, intuitively, meaningful. But let us remember a classical Piagetian question: In a box there are 20 wooden marbles, most of which are brown and a few of them white. A six year old child is asked whether there are more wooden or more brown marbles. The child generally answers that there are more brown marbles "because there are only a few white ones". The explanation of the child's difficulty is that the brown marbles belong simultaneously, in the above comparison, to *two different groups of objects*, and this is *practically* impossible.

According to Piaget, at the concrete operational level (after the age of 7—8) children become able to solve the problem but only if it is presented in concrete terms (that is, if the child *sees* the marbles and decides that the whole is bigger than a part of it). A similar problem presented in formal terms remains unsolved until the emergence of the formal operational stage.

Actually, the difficulty of mentally duplicating a number of elements of a set for the sake of a comparison may not have an impact on a certain category of problems but may continue to be active in respect of others. The result is a strange intuitive obstacle in solving some categories of problems. One known instance refers to calculating probabilities. One of the items in a probability questionnaire asked 5th, 6th and 7th grade pupils to solve the following problem. "In a box there are 3 red and 4 black marbles. One extracts a marble by chance (without looking) and one has to calculate the probability for extracting: a red marble; a black marble; a red marble after extracting a black marble etc." In what follows we refer only to the 5th grade children. The first question was solved by 6.7%, the second by 30.9% and the last by 21.2%. All these subjects had attended an elementary course in probability. With regard to the first two questions, the most frequent error consisted in comparing the two sub-sets instead of relating the "expected" set to the whole set of possible outcomes. A second typical error was the following. In a problem in which an extracted marble is not replaced, the child forgets that one has to diminish by 1 *both* the number of expected outcomes and the number of possible outcomes (Fischbein and Gazit, 1984).

In short, in this case, the obstacle consists in the intuitive difficulty of comparing the "part" with the "whole", an operation which requires the act of considering a part of the elements twice. The capacity to perform that "duplicating" operation — which appears at the age of 7 when checked in very simple, concrete situations — is still absent at the age of 10 when requested in more complex, formally presented, conditions.

A different type of example is analyzed by Raymond Duval (1983). Among other questions, he asked 12—13 year old subjects to compare the set of natural numbers with the set of even numbers.

According to Duval, many of his subjects were facing a genuine difficulty when trying to answer because they discovered that the numbers belonging to the set of even numbers belong also to the set of natural numbers. "Six is, at the same time, an element of *N* — the successor of 5 — and an even number". Some of the subjects were puzzled by this situation and consequently were not able to perform the comparison. Let me quote from one of the protocols:

Phillipe: Are there more whole numbers? . . . Sure, obviously. Oh, no! It depends if the whole numbers include the odd ones.
R: Yes the whole numbers include the even and the odd ones as well. All right.
Phillipe: But this is silly. There are no more whole numbers than even numbers because if one takes away the even numbers which one has not the right to use anymore, only the odd numbers remain. (Duval, 1983, page 406, my translation.)

According to Duval, the subjects cannot accept the bijection of the set of natural numbers with the set of even numbers because, in order to perform such a comparison, one must admit that the same element may have a double existence.

The phenomenon described by Duval, *the duplication obstacle*, represents a real intuitive obstacle for some formal types of reasoning. It has also been identified in other mathematical domains. Having to solve a geometrical problem one has sometimes to consider the same element — a side, an angle, a vertex — more than once, as if the respective element has a double, a triple etc. existence. A good example of this is an elegant proof (known to Pappus) of the theorem that if two of the angles of a triangle are equal, then two of the sides are equal. This proof goes as follows:

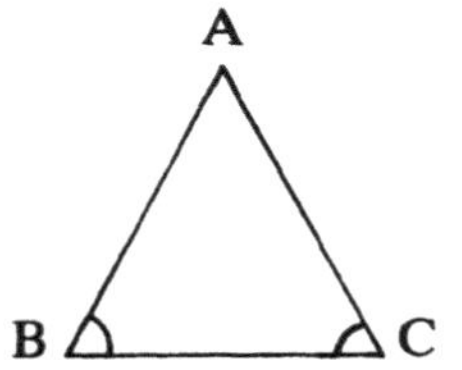

Fig. 1b.

Consider triangles *ABC*, *ACB*.

$\Delta ABC = \Delta ACB$ (given)
$\Delta ACB = \Delta ABC$ (given)
$BC = CB$
$\therefore$ triangles *ABC*, *ACB* are congruent
$\therefore AB = AC$

This proof depends on seeing a "single" triangle as two "different" triangles. It is well known that children may face difficulties when the *same* element plays simultaneously *different* roles in different figural structures.

But, with regard specifically to the comparison between an infinite set and one of its proper subsets, I think that the "duplication obstacle" plays a less important role than that assigned to it by Duval. I do not deny that, sometimes, a subject may be disturbed by the fact that, in performing the comparison, there are elements which intervene twice. But it seems to me that the equivalence of a set with one of its proper subsets represents a very disturbing situation by itself. In the protocol quoted above, Phillipe affirms, first, that the set of whole numbers is evidently bigger than the set of even numbers. But he is troubled by a previous discussion which has led to the idea that the set of whole numbers is equivalent to the set of squares of whole numbers ("each number has its square"). *He feels the contradiction and cannot overcome it.* Only at this moment, puzzled by the whole story, he somehow regresses to a more primitive difficulty "the duplication obstacle".

I cannot imagine that a twelve-year-old child asked to compare the number of whole numbers in the interval (11—29), with the number of even numbers in the same interval, will not be able to make the comparison. I expect that he would say that this is a silly question.

The duplication obstacle represents a real intuitive difficulty especially in pre-operational and in early operational children. The same type of obstacle may appear again in older subjects in relation to some formal complex tasks. But with regard to infinite cardinals, I think that it plays only a secondary role.

Anyway, the duplication obstacle described by Duval is a real source of difficulty. It is a particular expression of the practicality of intuitive representations. Sometimes it appears even in subjects who have already acquired the mental prerequisites which should, in principle, enable their reasoning to perform at a formal level.

SUMMARY

Experience plays a fundamental role in shaping intuitions. In relatively constant conditions, it produces, in the long run, stable systems of representations implying structured programs of actions and expectations. At the same time experience tends to bias our primary intuitive interpretations because of its immanent constraints: the terrestrial (general and specific) conditions, and the practical, concrete nature of every behaviorally meaningful activity (finiteness, concreteness, impossibility of ubiquity).

In our opinion, new intuitive attitudes can never be produced by mere verbal learning. Verbal explanations may enrich the child's ideas, the child's information about a certain reality. It may help him to understand and logically justify statements taught during school instruction. But the kind of specific belief, the subjective acceptance of ideas and representations as intrinsically valid, which characterize intuitions, can be attained only as an effect of direct, experiential involvement of the subject in a practical or mental activity. This requirement is perhaps not always achievable in science and mathematics education. Many statements have to be accepted formally as imposed by formal constraints. Most of the Cantorian statements, for instance, are in that category. But also many ideas in modern physics have to be accepted via mathematical symbolism (in the theory of relativity, in quantum physics etc.) without an adequate intuitive support. But similar situations may also be encountered in elementary mathematics or in classical physics. There are no intuitive justifications for relations like $a^0 = 1$ $(-a) \cdot (-b) = ab$; $(-a/b) \cdot (-c/d) = (ac/bd)$, or for the statement that a body will move indefinitely if no force intervenes. It is impossible to imagine behavioral or experimental conditions which would directly support such propositions.

The general didactical recommendations are, then:

1. It is important to deepen the student's intuitive understanding of the various concepts and statements. This can be done only by creating didactical situations which would require a personal, experiential involvement of the student's mental productive activity in the respective domain. We consider, for instance, that a successful teaching of statistics and probability cannot be attained by teaching theorems and solution procedures only. The student has to experience, first practically (and of course mentally) operations with dice, coins and marbles for watching, recording and summing up different sets of outcomes. He has to *live* the fascinating experience of situations in which individually unpredictable outcomes tend, when considered as mass phenomena, to produce certain structures, to display certain regularities. A student who does not first live such an experience will never *live* (understand and manipulate intuitively) statistical and probability statements. He will remain confined within the limitations of standard questions. I have chosen this domain as a paradigm mainly because it parallels, in a very

interesting and surprising manner, the domain of space intuitions, although in the former case it is the *uncertain* and in the latter the *certain* which predominates. In both, experience plays a fundamental role in shaping programmes of action and systems of expectations organized in intuitions.

2. On the other hand, sometimes, such practical, behaviorally meaningful situations cannot be generated with regard to scientific or mathematical notions in a natural, direct manner. In such cases one has to refrain from forcing artificial, sophisticated examples upon the student. The student has to learn that in science and in mathematics not everything is intuitively understandable, visually or behaviorally representable, that many statements express logical implications of generalisations going beyond the limited possibilities offered by the empirical, common conditions of our terrestrial life. If there is an intuition to be created here it is the intuition of the *non-intuitive*, the intuitive understanding of the fact that many conceptions are by their very nature beyond our intuitive capabilities, although rationally valid. Such an intuitive understanding is also attainable by experience — the experience of the non-representable although intellectually manipulable notion. One lives the conflict and the displeasure, one lives the effort to overcome the conflict, one lives, finally, the acceptance as clear and intellectually consistent of the particular statement or notion. *Such an intuition expressed in accepting the non-intuitive as meaningful on logical grounds represents a fundamental acquisition of science and mathematics education.*

CHAPTER 8

THE PRACTICALITY OF INTUITIVE MEANINGS, ANALYSIS OF AN EXAMPLE: THE NEGATIVE NUMBERS*

It took 1500 years until the mathematicians got used to the "rule of signs" of the directed numbers (Glaeser, 1981, p. 303).

The main obstacle consists in the fact that the concept of a negative number contradicted the concept of number itself as it had originally been developed in the history of mathematical reasoning. A negative number is a counter-intuitive concept because it apparently contradicts the notion of existence itself — if existence is considered with its practical meaning. But practicality seems to be one of the fundamental attributes of an intuitively acceptable notion. Certainly one may consider, conventionally, that a debt is opposite to possession, that moving leftward is the opposite of moving to the right, etc. But in all these instances one refers, in fact, to real magnitudes which may be expressed by numbers having the same practical meaning as the "positive" ones. The negative numbers appeared in the history of mathematics as a kind of artifact, as by-products of improperly designed mathematical problems.

It is true that Diophantus of Alexandria (end of the 3rd Century A.D.), in his 4th book on Arithmetic, mentions the sign rule of directed numbers. But this rule appears to Diophantus only as a transitional procedure in order to get an "acceptable" number, i.e. a positive one.

As Glaeser remarks "The clandestine" use of directed numbers has preceded their *understanding* by 1600 years! (Glaeser, 1981, p. 314).

According to Simon Stevin (1540—1620) a number expresses the quantity of a thing. He frequently uses negative numbers in his calculations but only as transitory expedients. Fermat (1601—1665) mentions various procedures for obtaining "acceptable" roots when "false" (negative) roots appear. In *The Mathematical Dictionary* published by Jacques Ozanon (1691) one may find under the heading "roots", the categories: Genuine, false and imaginary numbers. "A false root is the negated value of the unknown letter" (*cf.* Glaeser, page 316).

What is of great psychological relevance to our discussion is that the resistance shown to accepting the negative numbers did not come only from non-professionals. Mathematicians who had no objection to the use of other mathematical abstractions, for instance "non-material" entities (like points or lines), were opposed to conferring a formal mathematical status on the

* Most of the historical information used in this chapter has been taken from the excellent paper of G. Glaeser (1981).

negative numbers. The negative numbers have been mentioned and used since antiquity but they have not been accepted as being, by themselves, conceptually meaningful entities. For many centuries attempts have been made to attribute to the negative numbers some kind of a practical, behavioral validity in order to solve the problem of their legitimacy.

Descartes himself was concerned with finding a procedure to eliminate the negative roots of equations (by changing the point of origin).

The Scottish mathematician MacLaurin (1698—1746), seems clearly to have understood the formal nature of mathematical entities. "It is not necessary", he wrote in 1742, "to really describe the objects of our theories or that they should really exist. But it is essential that their relationships should be conceived clearly and deduced obviously." (*Cf.* Glaeser, 1981, page 318.)

Nevertheless, the same MacLaurin will write, later on in his *Treatise of Algebra* (1748) that "an *isolated quantity cannot be considered as being negative*: it may be so only by comparison. A negative quantity is not, rigorously speaking, less than nothing. It is not less real than a positive quantity, considered in an opposite sense." (*Cf.* Glaeser, *ibid.*, page 117.)

Although he has explicitly stated, as a general principle, that mathematics is a science of formal relationships and not of practically existing objects, he was not able, when considering a specific type of mathematical entity to liberate himself from the need to confer on these entities a practical, behavioral meaning. His model was that of practical operations with quantities existing empirically. ". . . for instance", he writes, "the value of money somebody expects to receive and that he owes; a line drawn to the right and a line drawn to the left; the altitude above the horizon and the depth beneath it." (*Cf.* Glaeser, p. 317.)

All these are not to MacLaurin mere didactical examples. *They are intended to justify the use of the concept of negative magnitudes.* He was not able to reach a point from which he could declare: The negative numbers *have* a formal existence, axiomatically justified and defined in the structure of mathematics. On the contrary: *he denies* their "absolute" existence simply because he cannot identify a real phenomenon which would be less than nothing! One may assume that even the term "relative numbers", as used in the French terminology, is in fact a survival of this old concept that negative numbers do not exist "on their own" but only as a kind of symmetrical, virtual mirroring of real quantities.

The difficulty becomes still harder when referring to operations with directed numbers. Let me refer specifically to the operation of multiplication. These difficulties somehow parallel those encountered with decimals.

One may intuitively conceive a situation in which a positive and a negative number are multiplied, but only if, according to the problem, the operator is represented by the positive number. Three times (−4) may be intuitively perceived, or $(-4) + (-4) + (-4) = -12$. And this means, intuitively, that, for instance, by borrowing three times 4 dollars you have, finally, a debt of 12 dollars. By contrast, (-3) times 4 has no intuitive meaning.

It is the same with decimals. While 3 times 0.65 means intuitively $0.65 + 0.65 + 0.65$, 0.65 times 3 (with 0.65 as the operator) has no intuitive meaning.

The acceptance of $(+a) \times (-b)$ as *always* being equal to $-ab$ (no matter which is the operator and which is the operand) is based on the law of commutativity; that is, through a shift to the formal level.

But the problem of $(-a) \times (-b)$ is much harder, not because, formally, it presents a special difficulty *but simply because even very fine mathematical minds could not, for a very long time, completely rid themselves of the impact of implicit intuitive models.*

For instance, d'Alembert (1717—1783) referring to the term *Negative* in the Encyclopedia writes that "The simple and natural enunciation of the problem of $(-a) \times (-b)$ should be to multiply $(+a) \times (+b)$ and thus to get $+ab$ (*Cf.* Glaeser, p. 324). And the well known mathematician of his time, Lazare Carnot (1753—1823), writes, referring to operations with negative numbers: "A multitude of paradoxes and relative absurdities result from the same notion; for instance, that -3 would be smaller than 2; nevertheless $(-3)^2$ would be greater than $(2)^2$, that is to say that, considering two unequal quantities, the square of the greater would be smaller than the square of the smaller: This shocks every clear idea which one may get referring to the notion of quantity" (*Cf.* Glaeser, p. 326).

It seems obvious that, for Carnot, the notion of number remained tacitly linked to that of a concrete magnitude. The operations with numbers were in fact to him *practical manipulations of concrete magnitudes.*

As Glaeser put it: "Let us remember that Lazare Carnot was a member of the Academy of Science. He was not a child who failed his examinations."

After hundreds of years of unsuccessful or only partially successful attempts (MacLaurin, Euler, Laplace, Cauchy, etc.), a German mathematician, Herman Hankel, finally solved the problem. In his *Theory of Complex Numbers* (1867), Hankel shows a complete change of perspective. He no longer tries to find concrete models for justifying the negative numbers. To him, the negative numbers are not symbols of given realities but, rather, formal constructs. Hankel considers the additive properties of the set of real numbers and the multiplication properties of the set of positive real numbers. He proposes to extend the multiplication in $\mathbf{R}^+$ to the multiplication in $\mathbf{R}$ by respecting a permanence principle: The structure obtained must be algebraically consistent. He states the following theorem: The only multiplication in $\mathbf{R}$ which may be considered as an extension of the usual multiplication in $\mathbf{R}^+$ by *respecting the law of distributivity* to the left and to the right is that which conforms to the rule of signs. His proof is very simple and reminds us of that of MacLaurin (*Cf.* Glaeser, pp. 337—338). And Glaeser concludes: "The revolution accomplished by Hankel was *to refuse to look for a good model*" (*ibid.*, 343).

According to Glaeser, the case of the negative numbers is only one example among many other instances of the struggle of mathematical reasoning to free itself from its enslavement to concrete empirical constraints.

When, in the course of the history of sciences, the formal game or the experience revealed some intellectual object which was unacceptable to the natural way of thinking, one started to look for a good model which could have appeared as familiar in the respective time. Thus, the invention of the non-Euclidean geometries became acceptable only when Beltrami, Gauss and Poincaré proposed some representations which were intellectually manageable. (Glaeser, p. 340, my translation.)

The difficulty of accepting the negative numbers as meaningful mathematical entities derives from the difficulty of identifying a good intuitive, familiar model which would consistently satisfy *all* the algebraic properties of these numbers, says Glaeser. As a matter of fact, such a model does not exist. One may create some models, but only by using a system of artificial conventions.

For instance: one may represent positive entities on a number line by distances to the right, negative ones by distances to the left, with the point 0 symbolizing the observer's location. Analogically, in respect to time: a future time is positive, a past time is negative, zero represents a present instant. A similar symbolization is used for indicating the sense of speed. If a car moves to the right, after a certain (positive) interval it will have travelled a certain, conventionally positive, distance.

But one may imagine a body which travels to the left (the direction of motion being marked (−)). What about the distance, measured from the point 0, at which the car was *in the past* (a negative time)? In the past, the car was *to the right* with regard to the observer and this means, according to the above convention, a positive distance. Thus, by multiplying two negative numbers (a negative time and a negative speed) one gets a positive number, that is to say a distance on the right.

Such a model is certainly very interesting, but it has an important drawback: it makes neither the concept of negative numbers nor the "rule of signs" of the directed numbers intuitively more acceptable. On the contrary, it only makes them more obscure, because one has built a model to fit, step by step, through a system of artificial conventions, the "rule of signs" of the directed numbers.

The main cognitive task of a model is to produce the kind of obviousness and self-consistency necessary for an intuitive acceptance of a concept or an intuitive leap in the course of a problem solving process. This is not the case with the above model. It is too sophisticated, it requires too many artificially connected conventions. Its real task is to prove that one may identify situations to which the rule of signs may apply in a noncontradictory manner.

In our opinion there are two essential intuitive obstacles concerning the negative numbers. One is the concept itself which is, as I said, intuitively contradictory. The notion of *quantity* has a naturally practical connotation: it is intrinsically something more than nothing. The concept of number is naturally related to this practical meaning of quantity. The notion of negative number is, then, practically inconceivable and this fact has reappeared, time and again, in the history of mathematics. Moreover: there is no *practical* requirement to invent negative numbers. There are no practically meaningful magnitudes the symbolization of which would absolutely require the use of negative numbers. On the contrary: things appear much more intelligible when using only non-directed numbers. Why should we say that somebody possesses −$5 if one may say that he has a debt of $5? Why should one say that the temperature is −5 and not 268 in Kelvin units or, simply, 5 degrees below zero, or 5 degrees cold? The negative numbers are a mathematical creation like the imaginary numbers: they are a by-product of mathematical calculations and not the symbolic expression of *existing properties*. If one encounters the expression $a/f(x)$, one hastens to remind oneself that for $f(x) = 0$ it has no meaning. In principle the same could have happened with the negative numbers. When they appear as the roots of an equation, one can discard them as being meaningless (that is to say as *practically* meaningless).

In order to accept them, one has to take a fundamentally different view. One has to admit that practicality is *not* a criterion in accepting an entity as being mathematically valid. The mathematical validity of a concept is based on the possibility of defining it and of operating with it consistently in the realm of a certain axiomatic structure.

By assigning the "+" and "−" signs to numbers one gets what has been called *directed numbers*. Formally, one can operate consistently with that new, enlarged, set. Moreover: new, interesting properties emerge, as for instance the group structure of the set of directed numbers with addition as the law of composition and zero as the neutral element. Such a fundamental transformation, prepared by centuries of theoretical debates, did indeed take place in the 19th century and has definitively established the legitimacy of the negative numbers.

Nowadays, pupils get used to the concept of directed numbers quite early and, therefore, the intuitive difficulty is somehow weakened. Besides, pupils get used to some simple models which may be consistently applied to the additive properties of this class of numbers; for instance, the number line on which the two opposite signs are represented either by magnitudes considered to the right and to the left of the origin or by displacement of a point to the right and to the left, respectively.

A second difficulty appears when considering multiplication and division problems. We refer to the situation in which the negative number plays the role of the operator. As long as one has to do with a multiplication of a positive and a negative number, it may be interpreted as a repeated addition with the positive number playing the role of the operator: $(-2) \times (+3) = (-2) + (-2) + (-2) = -6$. But if one refers to the multiplication of two negative numbers, the repeated addition model is of no use: a negative operator has no intuitive meaning. Whereas one may consider $(+3) \times (-2) = (-2) + (-2) + (-2)$, one cannot find an intuitive interpretation according to which a magnitude is considered (-3) times. One needs new conventions for this and this makes the whole story intuitively indigestible.

I agree, then, with Freudenthal, who claims that the chapter of negative numbers has to be treated formally from the beginning. In his view this is the first opportunity offered to a pupil to consider mathematical concepts from a formal deductive view point. "I think", he writes, "that the need for a rationalization exceeding intuitivity is first felt with negative numbers." (Freudenthal, 1973, p. 280.) He suggests that even the number line support be abandoned and that we should use what he calls "the inductive—extrapolation method". Various exercises are suggested which would compel the learner to accept, for the sake of consistency, the rule of signs (Freudenthal, 1973, p. 281). "The formal treatment of this topic", he writes, "foreshadows what on a higher level becomes a rigorous proof." (Freudenthal, *ibid.*, p. 282.)

CHAPTER 9

FACTORS OF IMMEDIACY

In this chapter, various aspects of the immediacy characteristically associated with intuitive cognitions are considered, namely (a) visualization, (b) availability, (c) anchoring, and (d) representativeness.

VISUALIZATION

This is the main factor contributing to the production of the effect of immediacy. Its role is so important that very often intuitive knowledge is identified with visual representation. It is a trivial affirmation that one tends naturally to think in terms of visual images and that what one cannot imagine visually is difficult to realize mentally. Poincaré referred to a class of intuitive mathematicians, whom he termed "geometers". Hilbert, in describing the ways in which a mathematician thinks, reminds us of the fundamental role of images. Such examples are well known. It is clear that the process of structuring visual representations is governed by its own laws, as described by Gestalt psychology (the laws of "Pragnanz", of similarity, of proximity, of closure, of good continuation). These laws may affect the process of intuitivization itself. It may be assumed that their importance for the theory of intuition exceeds the direct implications for perceptual knowledge. However, in the absence of experimental evidence the matter will not be discussed here.

When envisaging the role of images in structuring intuitions it is worth keeping in mind that visual representations are *not* by themselves intuitive knowledge. Visual images are an important factor in immediacy, but immediacy is not a sufficient condition for producing the specific structure of an intuitive cognition. By simply perceiving the schema of an electronic device one does not get a deep, direct understanding of how the device works (unless you have been specially trained). If a pupil sees the famous "figural proof" of the Pythagoras theorem this fact does not, by itself, help him to get an intuitive understanding of the theorem and the proof.

Moreover, images as models may inject into the related conceptual process properties and relationships which do *not* belong properly to the conceptual structure (points as spots, lines as strips etc.) and this may disturb the reasoning process itself.

Nevertheless, visualization, embedded *in an adequate cognitive activity*, remains an essential factor contributing to an intuitive understanding.

Let me quote Roger Shepard in this regard:

Fig. 2.

The effectiveness of non-verbal processes of mental imagery and spatial visualization . . . can perhaps be explained, at least in part, by reference to the following interrelated aspects of such processes: their private and therefore not socially, conventionally or institutionally controlled nature; their richly concrete and isomorphic structure; their engagement of highly developed, innate mechanisms of spatial intuition; and their direct emotional impact. (Shepard, 1978, p. 156.)

Let me also add that visual representations contribute to the organization of information in synoptic representations and thus constitute an important factor of globalization. On the other hand, the concreteness of visual images is an essential factor for creating the feeling of self-evidence and immediacy. A visual image not only organizes the data at hand in meaningful structures but it is also an important factor guiding the *analytical* development of a solution; visual representations are an essential anticipatory device.

Shepard also emphasizes another fundamental contribution of visual imagery which may be related to intuitive cognition — the non-conventional, personal, subjective, even emotional roots of mental imagery. The term immediacy then gets a new dimension. Immediacy means not only that the given reality is directly perceived, but also that the individual is directly, personally, somehow emotionally, involved in the given reality.

Intuition, as we have frequently emphasized, implies a kind of empathy, a kind of cognition through direct, internal identification with a phenomenon. A visual representation with its rich, concrete details mediates such a personal involvement, generally much better than a concept, or a formal description.

Moreover, as Shepard has emphasized, visual representations and, in general, mental imagery play a considerable role in creative activity — science, mathematics, art etc.

From the rich bibliography used by Shepard let me recall one example, that of Friedrich Kekule, the man who discovered the hexagonal structure of the benzene molecule and with this thereby revolutionized organic chemistry. Trying to concentrate on the problem he had been working on for years, one afternoon in 1865 he imagined the atoms that, as he wrote,

> ... were juggling before my eyes ... my mind's eye sharpened, by repeated sights of a similar kind, could now distinguish larger structures of different forms and in long chains, many of them close together; everything was moving in a snake-like and twisting manner. Suddenly, what was this? One of the snakes got hold of his own tail and the whole structure was mockingly twisting in front of my eyes. As if struck by lightning, I awoke ... (quoted by Shepard, p. 147).

This was the idea — the ringlike, closed structure of the arrangement of the atoms in a benzene molecule!

We call this an anticipatory intuition, because it anticipated the full analytical solution.

Visualizing does not only mean to "see" mentally. The image is a dynamic, constructive representation; in the above example the scientist was playing with the image, with the chains of atoms.

Let me now recall another example (this time from mathematics) quoted by Poincaré. In his book *The Value of Science*, Poincaré devoted a chapter to 'Intuition and Logic in Mathematics'. He made an attempt to classify mathematicians into two groups: those who think mainly in images (the geometers); and those he called the analysts, the pure conceptual thinkers. Poincaré mentions several examples of analysts and geometers and among those in the second category he cited the famous German mathematician Felix Klein.

> On the other hand [he writes] look at Professor Klein: he is studying one of the most abstract questions of the theory of functions to determine whether on a given Riemann surface there always exists a function admitting of given singularities. What does the celebrated German geometer do? He replaces his Riemann surface by a metallic surface whose electric conductivity varies according to certain laws. He connects two of its points with the two poles of a battery. The current, says he, must pass, and the distribution of this current on the surface will define a function whose singularities will be precisely those called for by the enunciation.
>
> Doubtless [continues Poincaré] Professor Klein well knows he has given here only a sketch: nevertheless he has not hesitated to publish it; and he would probably believe he finds in it, if not a rigorous demonstration, at least a kind of moral certainty. (Poincaré, 1920, p. 16.)

In Klein's imagery there is not simply a passive representation of a given reality. As in Kékulé's example, the German mathematician was experimenting with the representation he had imagined. In both cases the preliminary figural, global solution *although not yet fully developed*, was associated with a feeling of *intrinsic certainty*. The visual representation was more than an image; it was the intuitive solution to a problem in which the sensori-mental structure played an essential role.

In fact, it is not only visual images that help to structure intuitions — although they certainly are the most common form of imaginal support.

Sounds, in the case of musicians; muscular, motor and tactile representations, in the case of sculptors, etc. play an essential role in artistic creative activity.

In a discussion with Max Wertheimer — one of the founders of Gestalt psychology — Einstein once declared, referring to the creation of the theory of relativity: "These thoughts did not come in any verbal formulation. I very rarely think in words at all. A thought comes, and I may try to express it in words afterwards". (Wertheimer, 1961, p.228.)

Mental imagery is, in fact, a part of a more complex psychological domain discussed later in more detail, namely the domain of mental models.

AVAILABILITY

There are other factors, in addition to visualization, which may contribute to producing the effect of immediacy. Some of them have been described by Tversky and Kahneman in various papers, namely *availability, anchoring* and *representativeness.* With regard to mathematical thinking these notions have been mainly related to probability estimations but, in fact, these phenomena possess a much broader range of implications.

An example concerning availability mentioned by Tversky and Kahneman refers to the estimation of the relative frequency of various letters in an English text, in the first and in the third position of words. A typical problem was:

Consider the letter *r*. Is *r* more likely to appear in the first position? — the third position? (check one). My estimation for the ratio of these two values is . . .". (Tversky and Kahneman, 1982, p. 167.)

Five different letters were used. Each of these five letters was judged by a majority of subjects to be more frequent in the first than in the third position. These results were obtained despite the fact that all these letters were, in fact, more frequent in the third than in the first position.

The explanation of these findings is that it is *easier* to find instances of words beginning with a certain letter than instances in which the same letter appears in the third position. The frequency is estimated intuitively, considering not the real distribution (which, in fact, is *not* immediately available) but rather a distorted version of it, biased by availability.

The findings considered in the above example are not very interesting in themselves. Obviously, people asked to estimate *intuitively* the relative frequencies of letters located in different positions of words, will offer estimations based on easily available instances.

What is really interesting is that *people seem to take their estimations for granted.* They are not bothered by the fact that they have based their judgments, not on objectively representative samples, but on samples which may have been distorted by some uncontrolled factors. In other terms, *the important finding here is the influence of availability itself as a mechanism for producing intuitive solutions.* There are no objective data determining the choice in this case, but some subjective factors, the main role of which is to guarantee the immediacy and *the immediate acceptance* of the solution. The need for immediacy seems to be a stronger factor than the need for objective reliability.

Let me take another example. Subjects aged 10 to 15 were asked to estimate the number of permutations which may be produced with sets of 3, 4 and 5 letters. As the number of elements in the sets considered increased, the underestimation of the number of possible permutations also increased. The average estimation for 5 letters was about 16 while the correct answer is in fact 120. This phenomenon of condensation may be explained, at least partially, by the difficulty of producing a great number of distinct permutations. When their number is rather high, as an effect of the number of given elements, the relative frequency of available, effectively produced permutations (compared with the total number of possible permutations) decreases as the number of elements increases (Fischbein *et al.*, 1970, pp. 263—264).

Let me quote a different type of situation. One is asked to construct a tangent common to two given circles. On a first reading, the problem does not seem difficult. In fact, it is more difficult than it appears. Polya, who mentions the problem, suggests that the reader should consider, to begin with, a particular instance which would be easily accessible (Polya, 1954, p. 24). Normally, one tries instances immediately available, for example two circles with equal radii or two tangent circles. But no solving intuition is stimulated this way. In fact, one has to consider first the extreme case of a circle reduced to a point. Then the way to the solution becomes "visible". The difficulty with this problem is the fact that the particular instance from which one has to start (a circle reduced to a point) is not "available" in the sense of Tversky and Kahneman. The notion of a circle does not naturally suggest that of a point. Therefore, Polya's indication is not helpful in this case. On the contrary, it may produce a blocking effect. One insists on looking for a particular pair of circles and this, certainly, does not lead to a solution.

Availability is not necessarily a misleading heuristic. As Shelley Taylor pointed out: ". . . under some circumstances, use of the availability heuristic leads to perfectly appropriate conclusions" (Taylor, 1982, p. 199). In other situations, where there is a bias with respect to the information available, faulty conclusions may be drawn. Taylor cites such factors as biases in salience, biases in retrieval, and biases due to cognitive structures such as schemas, beliefs and values.

ANCHORING

Subjects asked to give a global, intuitive estimate of a quantity may be biased in their estimation by some "anchoring" facts.

Two groups of high school students were given 5 seconds to estimate the result of a multiplication operation.

The first group had to estimate the product 8 × 7 × 6 × 5 × 4 × 3 × 2 × 1. The second group estimated the product 1 × 2 × 3 × 4 × 5 × 6 × 7 × 8. The median estimate for the ascending sequence was 512, while the median estimate for the descending sequence was 2250. The correct answer is 40320 (Tversky and Kahneman, 1982, p. 15). Apart from the general underestimation, it is clear that, under the time constraint imposed, the size of the first few numbers encountered affected the overall estimate.

This example illustrates how an intuitive, global estimation is sometimes biased by the salient aspect (or role, or position) of a particular component of the data, and this is because the need for immediacy prevents the subjects from using all the given information.

Again, the interest of such findings is not the fact that the subject may be misled by some salient component. This is certainly trivial. What is surprising is the fact that he is ready, in certain circumstances, to ignore completely an important part of the relevant information for the sake of immediacy. Since the selection is automatic the subject not only does not use a part of the information, but he even ignores its importance for the validity of the judgment. The effect is overconfidence — characteristic of an intuitive approach.

REPRESENTATIVENESS

The representativeness effect may also be considered a factor of immediacy. Tversky and Kahneman relate representativeness especially to probability estimates. They describe this biasing factor in the following way:

> A person who follows this heuristic evaluates the probability of an uncertain event or a sample, by the degree to which it is (1) similar in essential properties to its parent population; and (2) reflects the salient features of the process by which it is generated ... in many situations an event A is judged more probable than an event B whenever A appears more representative than B. (Kahneman and Tversky, 1982, p. 33.)

Let us quote an example:

> On each round of a game, 20 marbles are distributed at random among five children Alan, Ben, Carl, Dan and Ed. Consider the following distributions:

I	II
Alan 4	Alan 4
Ben 4	Ben 4
Carl 5	Carl 4
Dan 4	Dan 4
Ed 3	Ed 4

> In many rounds of the game, will there be more results of type I or of type II? (Kahneman and Tversky *op. cit.*, p. 36.)

Distribution II is objectively more probable than distribution I. Nevertheless it has been found that a majority of the subjects questioned (36 out of 52) considered distribution I to be more probable.

The explanation of this result is that distribution I appears subjectively to be more representative of random allocation than the other. Distribution II appears to be too lawful to be a result of a random process.

As Kahneman and Tversky put it: "Subjects answer the above problems as if they ignored the individual nature of the two distributions, and compared, instead, the two respective classes of distributions" (Kahneman and Tversky *op. cit.*, p. 36). The representativeness bias is strong enough to nullify the impact of other factors on the subjects' estimations. Consider a second example.

A certain town is served by two hospitals. In the larger hospital about 45 babies are born each day and in the smaller hospital about 15 babies are born each day. As you know, about 50% of all babies are boys. The exact percentage of baby boys however varies from day to day. Sometimes it may be higher than 50%, sometimes lower.

For a period of one year each hospital recorded the days on which more than 60% of the babies were boys. Which hospital do you think recorded more such days?

— the larger hospital?
— the smaller hospital?
— about the same (i.e. within 5% of each other).

Out of 95 subjects 21 chose the first solution, 21 chose the second and 53 chose the third solution.

In fact, the expected number of days on which more than 60% of the babies are boys is much greater in the small hospital, because a large sample is less likely to deviate substantially from the theoretically expected value, which is 50%.

However, considering the two samples *equally representative*, subjects tend to disregard the role of the size of the samples for judging the likelihood of certain events.

SUMMARY

As has been repeatedly stressed, one of the characteristic properties of intuitive cognitions is immediacy. Visualization, whether mediated by an external representation or not, is very frequently involved — as stated earlier, what one cannot imagine visually is difficult to realize mentally. This is enshrined in the English language, in which, among the meanings of "to see" is "to discern mentally". Indeed, it has been pointed out that "seeing" is used more often in that sense than it is in the sense of "perceiving with the visual apparatus".

Tversky and Kahneman have analyzed a number of what they term "heuristics" which account for systematic biases in probabilistic judgments. In particular, they have shown that intuitive judgments are affected by biases in the samples of instances which are immediately accessible relative to the total relevant population of instances, coupled with ignorance or ignoring of these biases. Although they deal mainly with probabilistic judgments, the same phenomena are discernible in all types of intuitive judgment.

CHAPTER 10

FACTORS OF GLOBALITY

An intuitive judgment is expressed in a global view, in contrast to a logical inference which is analytical, discursive. Globalization is attained either by simply ignoring some of the components (and relying only on a few of them which may rapidly yield an apparently, coherent structure); or by organizing most of the available components into a kind of synthesizing, meaningful, unitary structure (mainly through a process of hierarchization).

The anchoring effect mentioned in the previous chapter may sometimes explain how the globalization effect is obtained, by processing only a part of the information and ignoring the rest. Children's reactions to Piagetian conservation problems such as those referring to quantity, length, number etc. may be accounted for in this way.

Piaget has called the phenomenon "centration". A child is asked to compare the mass of two equal balls of clay. After concluding that they are equal he is asked to stretch one of them into a sausage shape. The experimenter then asks the child whether in the two pieces — the ball and the sausage — there is the same amount of clay. Five year old children generally answer that the sausage contains more clay than the ball. This is a typical intuitive, global reaction, because the child does not yet possess the intellectual prerequisites for integrating all the information needed for a correct evaluation. He relies only on one aspect — the most striking one, which in this case is the length of the object (Piaget and Inhelder, 1941).

Of course, preoperational children answering in this way are not aware of the fact that they have extrapolated their global evaluation from only one dimension to the whole object.

Let me mention some other examples of global estimations based only on certain components.

In a series of experiments Siegler (1979) has shown that children, attempting to solve balance problems, proceed in a similar way before being able to consider both weight and distance. The subjects are presented with a balance scale to which are attached different weights situated at different distances from the fulcrum. The balance is immobilised by some mechanical means but the child has to predict whether the balance will stay in equilibrium or if one of its sides will go down when the balance is allowed to move freely.

Siegler affirms that, generally, children are consistent in applying certain tacit rules for predicting the behavior of the balance in different situations. Rule 1: the subject considers only the weight dimension; Rule 2: The subject considers the distance dimension but only when the weights are equal. He predicts correctly that the side with greater distance will go down. In other situations (different weights and different distances) the subject's predictions are in accordance with rule 1. Rule 3: the subject considers both distance and weight, except in conflictual situations in which the distance is greater in one side and the weight is greater in the other side. In this case the subject resorts to some undefined strategy. Rule 4: the moments for the two sides are compared.

Siegler found that rule 1 was predominant in 4—6 old children and rule 2 was frequent from 8 to 17 year old subjects. Rule 3 was predominant above 13 years of age. Rule 4 has been identified only in some adult subjects.

The main finding of this line of research is that most of the young subjects and many of the older ones seem to rely in their intuitive evaluations on only one component of the information provided (weight or distance), *while concluding globally and confidently for the whole situation.*

Wilkening and Anderson (1982) have contested Siegler's conclusion. In their opinion subjects usually tend to integrate the information obtained using some tacit simple algebraic rules like addition or multiplication.

Let me cite some examples. In order to check the findings relating to the balance scale, Wilkening and Anderson studied the same problem with a different methodology. Instead of asking the subjects to predict the effect of a given situation, they were asked to adjust weights and distances so as to equilibrate the balance.

It was found that only 6 year old children made predictions according to only one dimension. Above this age, subjects used integrative rules. At 9 years of age children used either additive or multiplicative patterns but for older children the multiplicative patterns became predominant.

An experiment relating to the evaluation of areas of rectangles yielded the following results. It was found that 5 year old children used an additive rule for evaluating areas (addition of the two dimensions of the rectangle) while adults consistently used the multiplicative rule. Some of the eight and eleven year old children seemed to use the additive rule whereas others of the same age-group used the multiplicative rule. Each subject seemed to be consistent in using his strategy (Wilkening, 1980, 54—58).

Let me add, with regard to the rectangle problem, that the subjects expressed their estimation of the areas by using a unidimensional evaluation scale.

An analogous study using probability tasks was performed by the same authors.

In a binary choice experiment, similar to the classical probability choice experiments of Piaget, the subjects were presented with two plates on which there were combinations of blue and yellow marbles. The subjects were asked to indicate the plate from which they would prefer to pick a marble with closed eyes in order to have a greater chance of getting a blue one.

In a numerical judgment task the subjects were presented on each trial with only one plate containing both blue and yellow marbles. They had to evaluate the likelihood of picking a blue marble by using "a happiness scale" (A graphic rating bar with a sad face at one end and a happy face at the other) (Wilkening and Anderson, 1984, pp. 21—26).

With the first technique (two plates) it was found that at least half of the subjects made their evaluations according to a single dimension (for instance, considering only the number of the blue marbles). The second technique revealed, on the contrary, that all the subjects were in fact integrating both categories of information (the number of blue and the number of yellow marbles).

It is worth mentioning the ingenious statistical procedure used by Wilkening and Anderson for processing their data. An Analysis of Variance design was applied. Each category of information possibly used by the subjects (for instance, number of expected outcomes and total number of possible outcomes) represented a main treatment for the design. If a certain type of information was used by a subject, the relevant main treatment appeared as statistically significant. Very interesting information was also obtained by considering the interactions. If the integrative procedure is addition the overall interaction term and all of its components should be close to zero. If the integration rule is a multiplicative one, the bilinear component of interaction should be non-zero, and the residual interaction should be close to zero (Wilkening and Anderson, 1982, p. 228).

A geometrical representation of the data corresponding to an additive model would consist in parallel lines while a multiplicative rule would yield a "fan" of nonparallel, divergent lines.

In all the above cited examples, Wilkening and Anderson asked the subjects to evaluate intuitively various magnitudes or to predict intuitively the outcome of a certain situation.

Their general conclusion was that subjects, even pre-operational ones, do not rely in their evaluations on only one component of the given situation (as is suggested by Siegler's interpretation) but tend to integrate in their representation a mixture of the basic components. The additive model seems to be the more primitive one (acting mainly in pre-operational and young operational children) while the multiplicative model becomes effective later on.

Let me emphasize again that the above presumed calculations are only inferred from statistical results — they are assumed to take place tacitly. *The subject's reaction is only a global intuitive one.* An important contribution of Siegler's, and Wilkening's and Anderson's studies is that they have found an analytical way of investigating intuitive reactions which by their very nature are direct and global i.e. apparently non-analysable. In order to interprete the above findings let us first return to the classical Piagetian conservation experiment. The child who compares the sausage-like bit of clay with the clay ball affirms that the sausage contains a greater amount of matter because, according to Piaget, the child considers only one dimension which becomes perceptively dominant. Wilkening and Anderson claim that, in fact, the child is able to integrate both dimensions in his evaluation but that a special technique is necessary in order to reveal the influence of these dimensions.

Let us consider a situation in which a five year old child has to compare two pieces of clay sausages having the same length but with different diameters. It would be very surprising if the child were to keep claiming that the sausages are the "same" quantitatively. He would certainly shift his attention from the length to the thickness dimension of the pieces of clay.

My standpoint is that one has to distinguish between non-integration and centration. It seems that neither the Piagetian school nor the Wilkening and Anderson interpretation takes this distinction into account.

If, for various reasons, a certain component does not seem to be relevant to the subject for serving a certain task, the effect of this component may not be detectable by the research methodology. This does not imply that the same component will not become active if the task is changed (for instance in the Wilkening and Anderson methodology compared with that of Siegler in the balance scale problem). Moreover, an apparently non-active factor, neutralized, for instance, by equalization, in fact takes part in the final cognitive decision just by its neutral information. The fact that the child is focusing his attention on one dimension (centration) does not necessarily imply that the other dimension has totally disappeared for him.

A six year old child is asked to choose from two boxes containing black and white marbles the box which gives a higher likelihood of extracting a white marble.

> In trial A the choice is between: (3 white; 5 black) and (4 white; 5 black).
> In trial B the choice is between (3 white; 5 black) and (4 white; 7 black).
> The child easily solves the first task but he is generally not able to solve the second.

Does this imply that in task A he compares only the white marbles while the black ones do not exist for him? Certainly not. He is able to make the correct choice only because he takes into account, tacitly, the fact that the black-marble factor is neutralized.

In some circumstances the extrapolation from, say, one dimension to the whole situation may be objectively justified by the neutrality of the other dimensions (via equalization). But it is also possible that some objectively relevant elements may be ignored simply because the subject does not possess the intellectual means for including them within a global evaluation.

The *extrapolation from a part to the whole seems to represent one of the main mechanisms of intuitive globalization*. At a first glance the above sentence seems to be a mere tautology. In fact, the main point here is not the extrapolation itself, but rather *the subject's belief that he evaluates the whole situation while in fact he takes into account only a part of the relevant information.*

The Piagetian studies have shown that after the age of 6—7 the child becomes able to "conserve" certain magnitudes like quantity of matter, length, cardinals. These reactions, too, are intuitive ones although Piaget uses the term *intuitive thinking* only for the pre-operational period (4—7 years of age). A seven-year-old child does not affirm that the ball and the sausage-like pieces of clay (in the above experiment) have the same amount of matter because the length compensates the width. If there is compensation, as Piaget believes, the child does not use it as an argument. He admits the equivalence directly, intuitively, without feeling the need for any justification. *This is a new intuition which develops in operational children as a part of their conservation capacities.*

Piaget generally uses, for describing the acquisitions of operational children, a logical, analytical terminology (inclusion of classes, compensation by reversibility of transformations etc.). Nevertheless most of the children's reactions during the concrete operational period described by Piaget (especially those related to conservation) are intuitive ones. In fact they remain intuitive all our life! It is self-evident that by changing the form of the piece of clay we alter neither the quantity nor the weight nor the volume of this object. The child acquires the conservation of weight at the age of nine and that of volume at the age of eleven. But he acquires them as new intuitions, as self-evident global answers:

In Piaget's terms: "... there always arrives a moment (between 6:6 and 7:8 years) when the child changes his attitudes: he does not need any reflection, he decides, he is even surprised that the problem is posed, he is sure about the conservation." (Piaget, 1967, p. 150.)

Both reactions are global ones, that which conserves and that which does not. Intuitions are developmental phenomena, and their structure changes as an effect of experience and general intellectual development.

According to Siegler, and Wilkening and Anderson *what changes and determines the transformation of intuitions are the implicit rules upon which the subject's decisions are based.* As I have already said, this finding is the main contribution of that approach. Such rules may be incorrect, or insufficient for making correct, cognitive decisions, but nevertheless they seem to guide systemtaically the subject's intuitive reactions. This is a fundamental finding for the theory of intuition because it supports the thesis of the lawfulness of the intuitive processing of information.

It seems also that the fact of relying on isolated factors in taking cognitive decisions is rather the exception than the rule.

People (including sometimes even pre-operational children) generally tend to integrate information according to certain integration rules (addition, multiplication or others). The degree of adequacy to the task of the integration rule adopted may vary according to age and instruction. Pre-operational children use an additive procedure for evaluating the magnitude of areas while older ones use multiplicative principles. But the fundamental finding is the fact itself that *intuitive evaluations are based more often than has previously been thought on consistent, tacit integrative procedures* (even before the subject is able to resort to correct, adequate ones).

Let me add a further example with regard to globalisation.

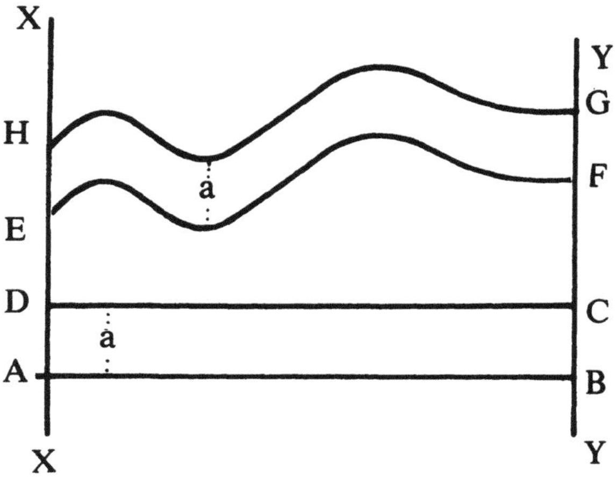

Fig. 3.

Consider two parallel axis *xx′* and *yy′*. Draw two parallel lines *AB* and *CD* perpendicular to the *xx′* and *yy′* axis. Let *a* be the constant distance between *AB* and *CD*. Draw two curves *EF* and *GH* (See figure 3) such that the distance between two corresponding points, measured on a line parallel to the *xx′* and *yy′* axes, remains constant and equal to *a*.

Subjects asked to compare the areas *ABCD* and *EFGH* affirm, generally, that the area *EFGH* is greater than the area *ABCD*. Asked to justify their answer people affirm that *EFGH* is "longer".

As a matter of fact the two areas are equivalent.

The common reaction — that the areas are not equivalent — recall the non-conservation reaction of pre-operational children in the clay-ball problem.

In the area problem, as in the clay-ball problem the subjects are focusing on only one dimension — the length — which becomes the dominant one. They do not consider attentively enough the width dimension. The conclusion, intuitively based, is that the two areas *as a whole* are not equivalent.

In other words, the globalization effect (in this case the whole judged on the basis of only one component) is not restricted to the domain of perception. It may be identified also at a conceptual level.

Intuitive globalization by omission referred to so far has been related to a process of extrapolation. The subject either simply ignores certain components because he is not able to integrate them in a global decision, or he ignores them because they are neutralized in the given circumstances.

But intuitive globalization may present also a completely different picture. This is the kind of globalization which takes place in what we have termed "conclusive intuition". After a correct solution of a problem or a correct proof to a theorem has been found, the solver tends to synthesize the analytical development in a global, meaningful, intuitively acceptable view. This global view may be expressed in verbal terms, in an image, in a gesture or in a combination of these.

In this case, it is by a process of hierarchical organization of meanings that the process takes place.

Let me quote the classical example of the theorem of Pythagoras: In a right triangle the sum of the squares of the two legs is equal to the square of the hypotenuse.

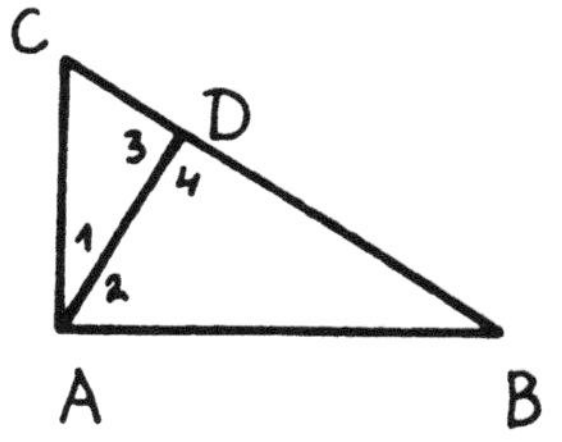

Fig. 4.

The complete proof of the theorem requires a number of steps. One has to prove first that $AC^2 = BC \cdot CD$ and then that $AB^2 = BC \cdot DB$. For this one considers first the triangles ADC and ABC and one proves that they are similar on the following grounds: $\sphericalangle A_1 = \sphericalangle B$, because their sides are respectively perpendicular, $\sphericalangle D_3 = \sphericalangle A$ (both are right angles). Being similar, the two triangles have proportional sides. And so on. It is a long chain of deductive steps. Considering the chain of steps only analytically, the pupil's understanding may easily get lost. An intuitive understanding of the proof would require the grasp of a hierarchical structure, the effect of condensation of the whole reasoning process into a global idea.

There are various ways of achieving this. One way is the usual one. Instead of the whole sequence of steps previously referred to, one emphasizes the main articulations. A first step: one proves that the square of each leg is equal to the product of the hypotenuse with the projection of the leg on the hypotenuse (using the similarity of the triangles involved). Second step: one adds the two expressions.

A much more elegant way is that described by Polya, which also may lead to a global, synthetical presentation. The main point is that one proves first the more general theorem: "If three similar polygons are described on three sides of a right triangle, the one described on the hypotenuse is equal in area to the sum of the two others" (Polya, 1954, p. 17).

This theorem is proved in the following way.

Let us consider figure 5.II. It is evident that the area of the triangle described on the hypothenuse is equivalent to the sum of the areas of the triangles described on the legs of the same right triangle. $A_1 = A_2 + A_3$.

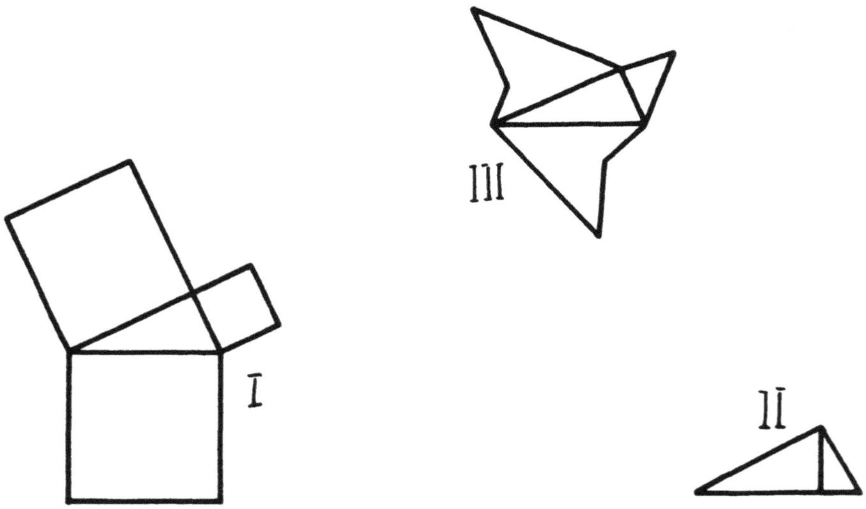

Fig. 5. (After Polya, 1954.)

Let us consider another triplet of similar polygons (figure 5.III). The angles of these polygons being built on the same sides *a, b, c* as the first triplet are proportional to the areas of the first triplet (the triangles *ABC, ABD* and *ADC* considered above). Their respective areas would then be

λA_1, λA_2, λA_3 and then one may write:
$\lambda A_1 = \lambda A_2 + \lambda A_3$ which leads to the general theorem mentioned above.

By applying this theorem to figure 5.I one gets

$$a^2 = b^2 + c^2$$

which is the theorem of Pythagoras. In Polya's terminology, this is an application by specialization of the general theorem. The quality of this proof is that it reveals the generality of the principle and, consequently, offers a deeper insight into the theorem itself.

The example of the theorem of Pythagoras emphasizes what has been said so far about the second type of globalization (that is to say, through a hierarchy of meanings and not by ignoring certain components as in the first type). The whole reasoning process is reorganized so as to emphasize the various conceptual levels of the reasoning structure — from those related to the main objectives to those representing only intermediary or subsidiary operations.

But the proof described by Polya reveals also another aspect with regard to intuitive thinking. That proof is certainly more difficult and more subtle than the usual one. Nevertheless, when genuinely understood, it is intuitively more convincing than the usual one. Its internal necessity is revealed to the subject in an unique synthetic grasp *despite* the fact that it is more difficult and more subtle.

The main reason, in my opinion, is that the proof described by Polya fulfils an essential requirement of an intuitive understanding. Beyond the theorem of Pythagoras referring to a special relationship between the sides of a right triangle, one grasps a more general property *which transforms the problem into a problem of figural compositions.* Through the directly convincing image of the triangle ABC composed of the two areas ADC and ABD, one "sees" the general theorem quoted above, which is simply obtained by multiplying each element of the relation referring to figure 5.III with the same factor. One obtains, then, a total fusion of the generality of a principle and a particular directly graspable (in this case figural) expression of it. *It is this kind of fusion which is the essence of intuition.*

When contemplating two intersecting lines, one sees that the opposite angles are equal. This is a matter of perception. It becomes a matter of intuition if, beyond the given image, one grasps the generality of the property as necessary and self-evident.

These examples seem, then, to suggest that intuitive globalization may, in certain circumstances, imply the fact of grasping, beyond the variety of particular instances, the generality of a principle seen as intrinsically necessary and self-evident.

Piaget, considering the various types of conservation, concluded that a child who is able to conserve a certain property (quantity, cardinality etc.) does not grasp the invariance of the property on the basis of explicit analytical arguments "The child does not need to think about it (il n'a plus a réfléchir) in order to make sure that the total quantity (of pearls) is conserved: he is certain a priori". (Piaget and Szeminska, 1964, p. 52.)

In our terminology, we would say that the child is intuitively convinced about the invariance of the given quantity despite the apparent changes of dimensions. Piaget claims that this intuitive conviction is based on the coordination of the transformations of the object's dimensions "concentrated in a unique act, instead of being effectuated step by step" (*ibid.*, p. 52).

SUMMARY

Let us summarize the various mechanisms of intuitive globalization identified so far:

1. By extrapolation, from a part to the whole. This may take place either by simply ignoring the remaining properties (the subject is not able to integrate all the elements in a unique comprehensive evaluation) or by neutralizing the rest of the properties on the basis of equivalences in the given situation.

2. By hierarchical organization: the subject grasps the elements involved in a situation as a hierarchy of properties (or objectives) in which a few global ones guide or justify the subsidiary ones. (For instance, the main idea of a solution justifying its successive steps.)

3. By grasping the particular relationships involved in a phenomenon as an expression of a general principle seen as a priori necessary and self-evident (For instance: changing the distances between elements in a given set does not change the cardinal because the cardinal is necessarily invariant — except when elements are added or subtracted).

4. A fourth mechanism, not specifically considered here, is certainly that of visualization. One of the main functions of pictorial representations in reasoning processes is that of producing a global, simultaneous, panoramic account of what is in reality a process, a succession of events.

Globalization does not lead necessarily to an intuitive acceptance but it may contribute to producing or enhancing an intuitive acceptance.

One may suppose that the effects of various globalization mechanisms are usually combined. For instance, in Polya's example with regard to the theorem of Pythagoras it is the general theorem which, in fact, helps the subject to organize hierarchically the various elements of the proof.

It is first the analogy between part I (the sum of squares) and part II (the sum of the triangles) which suggests the possibility of the general theorem (through analogy, via induction to the general idea). We then identify an act of globalization by neutralization and extrapolation. (The two instances I and II are considered to be equivalent and pointing to a common class of situations. In this specific case one extrapolates through induction.) One has, secondly, the general principle, the proof of which includes the intrinsic, intuitive necessity of the relation $A_1 = A_2 + A_3$ between the triangles of II. But it is only when this relation is seen not as referring to merely a particular construction but as the special case of a general principle (on grounds of proportionality) that the main idea of the proof is grasped intuitively (globalization by hierarchy). As already emphasized, what represents the core, the essence of an intuitive acceptance so beautifully represented in Polya's example is the fundamental unity between the particular, the specific, the directly convincing example and the general principle derived through similarity and proportionality from the particular case.

CHAPTER 11

INTUITION AND INTUITIVE MODELS

Models represent an essential tool for shaping intuitively acceptable cognitions. Whenever a person has to cope with a notion which is intuitively unacceptable, he tends to produce — sometimes deliberately, sometimes unconsciously — substitutes of that notion which are intuitively more accessible. Such substitutes are commonly called *intuitive models*.

Generally speaking, a system B represents a model of system A if, on the basis of a certain isomorphism, a description or a solution produced in terms of A may be reflected consistently in terms of B and vice versa (*cf.* Gentner, 1983).

CLASSIFICATION OF MODEL TYPES

Abstract Versus Intuitive

A first dichotomy distinguishes, roughly, between *abstract* and *intuitive* models. Mathematical relations (formulae, functions etc.) are usually *abstract models* of certain concrete realities. The quadratic function $s = \frac{1}{2}at^2$ is an abstract model for accelerated motion. Knowing a (the acceleration) one may determine the distance covered by a body in t units of time. The solution obtained in the abstract system is valid for the corresponding concrete phenomenon and represents an essential tool for predicting events with regard to that concrete phenomenon. *An intuitive model* is, by its very nature, of a sensorial kind. It may be perceived, represented or manipulated like any other concrete reality. For example, in order to represent vectorial magnitudes (like forces) oriented line segments are used. In order to represent directed numbers one uses the image of the number line with a conventional origin on it. In order to solve stochastic or combinatorial problems, one may resort to tree diagrams. The graph representing a function is also an example of an intuitive model. An intuitive model is not necessarily a direct reflection of a certain reality — very often it is based on an abstract interpretation of that reality. The graph of a function is an intuitive model of that function and the function, in turn, is the abstract model of a real phenomenon. For example, the phenomenon of a falling body → the quadratic function representing it → the graph (a visual, behavioral representation of the dynamic relationship between the variables involved). Intuitive models which use conventional, graphical means are generally called diagrams.

Explicit Versus Implicit

A second basic dichotomy distinguishes between *explicit* and *implicit* (or *tacit*) models. Sometimes models are *purposely* and consciously chosen or built, in order to facilitate the finding of a solution. One produces, for instance, maquettes or other types of simulation in order to study, in simplified conditions, the possible behavior of projected devices. Graphs, diagrams, histograms are deliberately drawn.

But, very often, models are produced automatically and used tacitly in connection with a certain reality. A person may be convinced that the object of his solving attempts is a certain phenomenon — the object of his interest — while his mental endeavors deal in fact with a model of it.

Analogical and Paradigmatic Models

A third classification, proposed by us, distinguishes between *analogical* and *paradigmatic* models.

Two entities are considered to be in a relation of analogy if there are some systematic similarities between them, which would entitle a person to assume the existence of other similarities as well.

In the case of analogical models, the model and the original belong to two distinct conceptual systems. In the case of a paradigmatic model, the original consists in a certain class of entities while the model is provided by an exemplar or a sub-class of the considered category.

One may establish an *analogy* between an electrical current and a flow of liquid through a very fine tube. The two phenomena belong to *two conceptually distinct classes.*

A child asked to identify the state of matter of a powder would usually be puzzled to learn that the powder is a solid. For him a solid should have *all* the properties of a compact object including its compactness (like a piece of wood or metal). Usually, to a child, a liquid is "a water". For example, the wick in a melted candle will not burn because the melted candle is "a water" (Stavy and Stachel, 1985, p. 9). Thus, water is a paradigmatic model for the class of liquids.

ROLE OF TACIT INTUITIVE MODELS

One of the hypotheses of this work is that tacit, intuitive models — both paradigmatic and analogical — play a fundamental role in every productive reasoning process. There is no productive reasoning activity without intuitive events consisting of globalization, concretization, extrapolation etc. Intuitive models are genuinely beneficial in respect of all these aspects. A model offers the solver a substitute of the original which, by its qualities, is better adapted to the nature of human thinking than the original. We think better with the perceptible, the practically manipulable, the familiar, the behaviorally controllable, the implicitly lawful, than with the abstract, the unrepresentable, the uncertain, the infinite. *The essential role of an intuitive model is, then, to constitute an intervening device between the intellectually inaccessible and the intellectually acceptable and manipulable.* The model has to code the data of the original (properties, processes, relationships), in its own specific, intuitively acceptable terms. The problem is solved in terms of the model and re-interpreted in terms of the original.

Conditions for Heuristic Efficacy of Models

This imposes some basic constraints on a model which, in practical conditions, may conflict amongst themselves. Firstly, *the model has to be faithful to the original on the basis of a structural isomorphism between them.* This implies that a set of invariant, consistent, coherent, reciprocal relationships must exist between the original and the model. A solution obtained in terms of the model and a solution obtained in corresponding terms of the original have to be equivalent. The validity of the first must be a sufficient condition for the validity of the second.

Secondly, the model has to enjoy a relative autonomy with respect to the original. A model which needs to resort to the original for generating and controlling each of its steps is heuristically useless. The autonomy of a dynamic system implies that it must be well structured, internally consistent, governed by its own laws.

Let us consider briefly an imaginary example. Imagine that, for attracting the interest of children, one attaches a certain image to each number from 1 to 20. One is a cat, two is a dog, three is a mouse, four is a cow, five is a bird etc. Addition is represented by a house in which two or more animals are put together. How much is two plus three? You cannot solve the problem unless you go back to "the original"; you add 2 + 3, you obtain 5 and then you may say that, according to the model, a dog plus a mouse yields a bird. The model, by itself, is completely incapable of solving any probem whatsoever when posed in terms of the original. It is absolutely unstructured. The effect of each combination of its terms cannot be predicted by considering only the model itself. One has to resort to the corresponding combination in the frame of the original for that purpose. *The heuristic capacity of a model depends on its autonomy in respect to the original. And such an autonomy, as already said, implies that the model must be able to function coherently on the basis of its own laws.*

Certainly, the above example is the product of an imaginary, extreme exaggeration, but one may find among the so-called "structured materials" used in mathematical education examples of such internally unstructured and consequently heuristically unproductive models.

Let me consider an opposite example in which two systems are completely isomorphic and yet each is autonomous with respect to the other. I am referring to numbers and geometrical figures. In principle, one may solve in algebraic terms every problem posed in geometrical terms and *vice versa.* If I want to find the locus determined by the motion of a point under certain constraints, I may, in principle, solve the problem using a geometrical construction. But it may be easier to solve the problem by algebraic procedures.

Using a model means *thinking productively* in terms of the model. A mental model is a reasoning device. It inspires solving strategies, suggests solutions and very often confirms the coherence, the meaningfulness of solutions obtained by means of the original. For all those purposes, the model has to enjoy a certain degree of autonomy.

A third condition for a model to be heuristically efficient is that it must correspond to human information-processing characteristics.

These three categories of conditions may lead to conflicting requirements.

The main advantage of an autonomous model is that the subject may rely on the model alone in order to solve various problems posed by the original. But, at the same time, this increases the danger that the solver may become so much captured by the model, that he may, sometimes, draw conclusions for the original from properties of the model which, in fact, are not relevant to the original. The danger is certainly much greater when tacit, uncontrolled models participate in a reasoning activity.

Consider these examples. Elementary particles are ordinarily represented as tiny balls. This helps in understanding the structure of the atom. But, at the same time, it hinders the acceptance of the wave properties of particles and *vice versa.* The common sense representation of radiation as a fluid which propagates under the form of waves makes it difficult to accept the corpuscular (quantum) character of energy. The fundamentally different intuitive models commonly attached to particles (matter) and radiation (energy) constitute a basic obstacle for accepting intuitively the idea that matter and energy are equivalent and that the fundamental components of reality have, *all of them*, a dual nature, namely they possess simultaneously corpuscular and wave characteristics.

Mathematicians know very well that a point is an abstract, nondimensional entity, and that a line is an abstract unidimensional entity. Nevertheless, various mathematical misconceptions in the history of mathematics may be explained by the intervention of illegitimate, uncontrolled pictorial models (an ink spot for a point, a narrow strip for a line). It is difficult to accept intuitively the equivalence between the sets of points of two line segments with different lengths. The main explanation is that points are usually represented as small dots. The dot model of a point is intended to help us to operate mentally with the non-representable concept of a point. Without such a dot model it would be difficult to understand propositions such as: "Two points determine a line" and "Two lines may have in common either one single point or an infinity of points." But the dot model may also suggest wrong interpretations because it may *tacitly* introduce in the reasoning process other properties of a spot which, in fact, do not belong to the *concept* of point, which are irrelevant to the original. An example is the statement that the sets of points of two line segments of different lengths *cannot* be equivalent. The longer line "tends" to contain more points. Or, another solution sometimes offered by children: "The points of the longer segment are bigger."

The usual tacit model corresponding to the concept of number is that of sets of points (in fact, sets of dots) as it is introduced from childhood. The model is very helpful for operating mentally with whole numbers (certainly small ones). One can see what it means to compare two numbers, what addition and subtraction mean etc. The set-of-spots model yields a spatialized, panoramic, behaviorally meaningful representation of number concepts and of operations with them (including the class and order properties of the concept of number). But the "set-of-dots" model contains also properties which, if made absolute, become obstacles to the enlargement of the concept of number. The concepts of rational and real numbers cannot be interpreted in terms of sets of discrete elements. The concepts of negative and complex numbers are not interpretable in terms of sets of dots. A dot has a material, spatial, discrete existence. These properties prevent it from being a reliable support for number concepts other than those of whole numbers.

It is very interesting that, usually, we are not aware of many of the properties of the models used in a reasoning process. Even if the model is used consciously there are properties which act as tacit components of it, as ground properties.

In certain normal conditions they do not influence the subject's interpretation at all, and then the model may be really useful. But in other conditions, such irrelevant components may intervene effectively, though tacitly, distorting, misleading, blocking the correct interpretation of the concept.

The dots representing points have color, form, magnitude. They are distributed in groups, which also may have different structures and shapes (rows, rectangles etc.). None of these properties are relevant to the formal concept of number. But each of them may affect, in a certain manner, the interpretation of the concept or a certain solving strategy, *without the subject being aware of the tacit intervention of these properties.*

SUMMARY

Let me summarize. A model has to be self-consistent and, at the same time consistent, on the one hand with the original, and on the other hand with features of human cognition. This situation imposes upon models a number of constraints which cannot easily be fulfilled simultaneously by the same mental device. A good model has to be an autonomous entity but at the same time a trusty mediator between the original situation and the solver's intellectual activity. Consequently, most of our tacit, intuitive models are imperfect mediators, leading often to incorrect or incomplete interpretations.

CHAPTER 12

MODELS AND ANALOGIES

THE ROLE OF ANALOGY IN MODEL CONSTRUCTION

Analogies are a very rich source of models. Two objects, two systems are said to be analogical if, on the basis of a certain partial similarity, one feels entitled to assume that the respective entities are similar in other respects as well. The difference between analogy and trivial similarity is that *analogy justifies plausible inferences.* A red house and a strawberry have the same colour but nobody will see any analogy between the red house and the strawberry. Analogy implies then, *similarity of structure*, a stock of common structured properties. (*Cf.* Gentner, 1983.)

As Gick and Holyoak have pointed out:

> The essence of analogical thinking is the transfer of knowledge from one situation to another by a process of *mapping* — finding a set of one-to-one correspondences (often incomplete) between aspects of one body of information and aspects of another. (Gick and Holyoak, 1983; page 2.)

Analogies become models if they play an effective role in interpreting or solving. As has already been mentioned, the difference between a paradigm and an analogy is that, in the first case, the model is provided by an example of the notion to be interpreted, while in the second case, one has to do with two different systems. The laws of the electrical current were first established by analogy with those of a fluid flowing through a very narrow pipe. In the case of the fluid one may consider the force applied to cause the fluid to flow, the speed of the flow, the rate of flow through a section of the pipe, and the resistance offered by the pipe (depending on its length and diameter). Analogically, with regard to the electrical current, one may consider a potential difference (or voltage) which corresponds to the force or the pressure applied on the liquid, the intensity of the current, which would correspond to the rate of flow, and the resistance offered by the conductor, which would correspond to the resistance determined by the features of the pipe. The famous law of Ohm, $V = IR$ is obtained by analogy with the law which governs the flow of a fluid through narrow pipes (Poiseuille).

In this example, considering some superficial similarities, one has tried to identify a group of common, structural properties. Certainly not all the properties are common. One has to establish the limits of the similarity. But the fact of identifying and using the similarities has provided the cognitive process with a fundamental stimulative support.

First, one gets a source of research hypotheses. Secondly — and this is in fact the fundamental quality of analogies — the model provides a compact, structured, relatively familiar, internally consistent mental object, a viable component of an active try-and-see reasoning process. If you contemplate the wire and you find that a cause A at one end determines an effect B at the other, it is practically impossible to think productively about the whole phenomenon without imagining some traveling, intervening agent. Therefore you think, normally, of a kind of fluid. And thus you obtain a concretely structured entity which inspires and guides your thoughts and which will be an essential component of a feedback monitoring process as well. You think about the model, you do not think of the abstract, as yet unstructured set of properties of the original (a travelling, as yet unidentified medium in this case). From the findings obtained by investigating the model, you infer possible properties of the original and you try to see if they fit. But basically the invention process proceeds by keeping the model in mind, the mental entity which is directly thinkable as a compact, self-consistent, intrinsically credible phenomenon.

Analogies are not always primarily based on superficial similarities. While trying to understand the structure of the atom, J. J. Thomson tried to picture it as a positively charged substance distributed more or less uniformly through the entire body of the atom, with negatively charged electrons embedded within it. But experimental findings suggested that the atom must consist of mobile particles. Rutherford proposed the planetary model, which was subsequently improved by Bohr, Sommerfeld, and others.

The Rutherford analogy was not an arbitrary choice. It was supposed to correspond to the then known facts in physics. It could better account for these facts than the previous, somehow static, model of J. J. Thomson. Thomson's model could not explain the fact that some of the particles, when shot in the direction of a thin gold foil, were repelled by the positive charges within the atom. Such an effect may be explained only if one considers that the entire mass of the atom is very small and is located close to its center.

The Rutherford model could be improved and adapted to new findings. The form of the orbits, the motion of the particles, the orbital radii etc., all of these properties could be related to new experimental findings.

An intuitive analogy helps to get a unitary iconic representation with a concrete behavioral meaning. An intuitive understanding then becomes possible. The reasoning process gets an "object", a representational system with its qualities of immediacy, globality, generativeness, intrinsic consistency and extrapolativeness.

N. R. Campbell in his *Physics* (Campbell, 1920) strongly pleads for the role of analogic models in developing a theory. Mary Hesse, who quotes Campbell, writes:

> A theory in its scientific context is not a static museum piece, but is always being extended and modified to account for new phenomena. (Hesse, 1966, p. 4.)

Campbell has shown that without the analogy, without a model, any such extension would be merely arbitrary. And Hesse quotes Campbell himself in continuation:

... analogies are not aids to the establishment of theories: they are an utterly essential part of theories without which theories would be completely valueless and unworthy of the name. It is often suggested that the analogy leads to the formulation of the theory, but once the theory is formulated, the analogy has served its purpose and may be removed or forgotten. Such a suggestion is absolutely false and perniciously misleading. (Campbell, 1920, p. 129; *cf.* Hesse, *ibid.*, p. 4.)

What Campbell affirms with regard to great scientific theories holds also for our current intuitions. As we have frequently stressed, in our view an intuition is, essentially, a theory (even if it is only a mini-theory). We commonly think in terms of models because they provide the process of reasoning with the structuring and stimulating ingredients necessary to its creative course.

ANALOGIC MODELS IN MATHEMATICS

Analogy frequently intervenes in mathematical reasoning. If a student knows that the area of a rectangle is $B \times I$, he may naturally extend the principle of this solution to the volume of a prism or a cylinder in which B becomes the area of the base of the prism or the cylinder. A similar analogical transfer is made from the triangle to the pyramid or the cone, from a trapezium to the frustrum of a pyramid or a cone. There is an analogy between the change of direction of a trajectory and the change of speed of a moving body. In both cases the rate of change is expressed by the derivatives of the respective functions.

Polya writes about "great analogies" in mathematics. He mentions the fundamental analogy between the domain of numbers and that of figures and that analogy represents the ground for analytical geometry. Polya also insists on the analogy between the finite and the infinite.

It is worthwhile distinguishing different types of analogies which may intervene, tacity or explicitly, in mathematical reasoning.

One may first consider two main categories of intramathematical analogies.

a. One, in which both the model and the original do not use explicit intuitive means but only a numerical-algebraic symbolism. Consider, for instance, the case of the operations with imaginary numbers, defined by analogy with real numbers, or operations with transfinite numbers, defined by analogy with finite cardinals.

b. In a second category of intramathematical analogies, there are those in which one term is an intuitive, usually a geometrical representation and the second term is a symbolic expression. The geometrical representations of functions based on the fundamental isomorphism between numbers and figures is the most important example of that category.

c. A third category of analogies which intervenes in mathematical reasoning is that in which the model is extramathematical, more specifically a material representation of the mathematical concepts. Structured materials (for instance those produced by Dienes or the Cuisenaire rods belong to that category). But one may also include in the same category the pictorial representations of numbers or of geometrical concepts: small spots for points, set of spots for numbers.

In a real reasoning process these types of models may, in fact, be mixed. Such mixtures of models are generally, components of tacit, uncontrolled processes and therefore may give raise to misinterpretations and to wrong solutions. The concept of a continuous function is associated with the notion of a continuous curve, the second being an analogic-geometrical model of the first. But, in turn, the ideal curve — which is supposed to correspond, ideally, to the function — is in fact imagined through a pictorial-model, a fine ink strip.

All these may be considered to be no more than psychological verbiage. As a matter of fact things are not so. The discovery made by Weierstrass of continuous functions defined over the reals which do not have derivatives at every point has shown the profound effect of such complex relationships. Because we cannot completely free, in our reasoning processes, the idea of continuity from its geometrical meaning and that meaning from its tacit pictorial representation, we are not intuitively ready to accept the notion that a continuous function (*alias* a continuous curve) may not have derivatives (or tangents, respectively) at all or at some of its points. A *drawn* continuous curve must intuitively have tangents indicating the slopes of the curve at various points of it.

Examples of Mathematical Analogies

Let me mention, in continuation, some examples of mathematical analogies.

(a) An abstract—abstract analogy
Polya has extensively discussed the role of analogy in mathematical reasoning. He quotes, for instance the history of the solution to the problem asking for the sum of the terms of the series:

$$1 + 1/4 + 1/9 + 1/25 + 1/36 + 1/49 \ldots.$$

Jacques Bernoulli had tried to solve the problem but he had not succeeded. It was Euler who solved it by resorting to an ingenious analogy.

The polynomial:

$$a_0 + a_1x + a_2x^2 + \cdots + a_nx^n.$$

can be represented as a product of n linear factors,

$$a_0\left(1-\frac{x}{\alpha_1}\right)\left(1-\frac{x}{\alpha_2}\right)\cdots\left(1-\frac{x}{\alpha_n}\right)$$

where α_1, $\alpha_2 \cdots \alpha_n$ are the roots of the equation obtained by equalizing the above polynomial with 0, ($a_n \neq 0$).

One may also consider a different variant. Let us consider the equation:

$$b_0 - b_1x^2 + b_2x^4 - \cdots + (-1)^n b_n x^{2n}.$$

Which has $2n$ distinct roots

$$\beta_1, -\beta_1, \beta_2, -\beta_2 \cdots \beta_n, -\beta_n.$$

We then may have:

$$b_0 - b_1x^2 + b_2x^4 - \cdots + (-1)^n b_n x^{2n} =$$

$$= b_0\left(1-\frac{x^2}{\beta_1^2}\right)\left(1-\frac{x^2}{\beta_2^2}\right)\cdots\left(1-\frac{x^2}{\beta_n^2}\right)$$

and (A)

$$b_1 = b_0\left(\frac{1}{\beta_1^2}+\frac{1}{\beta_2^2}+\cdots+\frac{1}{\beta_n^2}\right).$$

Euler considered the equation:

$$\sin x = 0$$

which yields, by expanding the left-hand side:

$$\frac{x}{1}-\frac{x^3}{1.2.3}+\frac{x^5}{1.2.3.4.5}-\frac{x^7}{1.2.3.4.5.6.7}+\cdots=0.$$

Let us consider this as an equation of infinite degree, the roots of which would be:

$$0, \pi, -\pi, 2\pi, -2\pi, 3\pi, -3\pi, \cdots$$

As Mark Steiner had pointed out, this was an inductive "leap" unjustified by the previous steps (Steiner, 1975, p. 103), a leap from a finite to an equation of an infinite degree.

In order to eliminate the root zero, Euler divides the left part of the equation by x and gets

$$1-\frac{x^2}{2.3}+\frac{x^4}{2.3.4.5}-\frac{x^6}{2.3.4.5.6.7}+\cdots=0$$

the roots of which are:

$$\pi, -\pi, 2\pi, -2\pi, 3\pi, -3\pi \cdots$$

By analogy with (A) one may conclude that:

$$\frac{\sin x}{x} = 1 - \frac{x^2}{2.3} + \frac{x^4}{2.3.4.5} - \frac{x^6}{2.3.4.5.6.7} + \cdots =$$

$$= \left(1 - \frac{x^2}{\pi^2}\right)\left(1 - \frac{x^2}{4\pi^2}\right)\left(1 - \frac{x^2}{9\pi^2}\right)\cdots$$

$$\frac{1}{2.3} = \frac{1}{\pi^2} + \frac{1}{4\pi^2} + \frac{1}{9\pi^2} + \cdots$$

$$1 + \frac{1}{4} + \frac{1}{9} + \frac{1}{16} + \cdots = \frac{\pi^2}{6}$$

(For more details see Polya, 1954, pp. 17—21).

As a matter of fact, Euler has extended the applicability of a rule from equations of finite degrees to an equation of an infinite degree.

This was an intuitive leap, based on analogy. Euler's conclusion was not at that time rigorously justified by a formal proof. As Polya has pointed out, the ground of the argumentation was an analogy, an analogy between the finite and the infinite. It was only ten years later that Euler gave a formal proof for his solution.

In the above example, the model used was an equation of finite degree and a relation between its roots and its coefficients. This is certainly not an intuitive model and one can scarcely see how the use of such a model would guarantee an intuitive (direct, apparently self-evident) interpretation of the final solution. One may concede that, in this case, intuition intervened in the moment of the transfer of acceptance — without proof — via extrapolation, from a finite set of elements to an infinite one. We do not possess a full introspective account of the discovery, and therefore there is no evidence that some genuinely intuitive model helped Euler to obtain the solution. Following Euler's reasoning, as it has been explicitly reported, one may suppose that the model was an intrinsically abstract one. If this is true, one may speculate that intuitions do not necessarily involve intuitive (pictorial) models.

An Intra-Mathematical, Figurative Analogy: The Geometrical Interpretation of Complex Numbers

It is well known that the "imaginary" roots obtained when solving certain algebraic equations have represented a puzzle for mathematicians. The term "imaginary" itself indicates the fact that these roots were considered out of the reach of a rationalistic view. They were not considered to be "true" numbers, but rather imaginary ones. And this despite the fact that the number concept, as such, does not necessarily in principle imply a realistic model. But, as a matter of fact, only in modern mathematics has a complete, formal interpretation of the operations with complex numbers been obtained by consideration of the formal definitions of addition and multiplication.

Before such a formal interpretation was given, mathematicians had invented an intuitively acceptable representation. On one hand, the internal consistency of such an interpretation proves that the original — the system of complex numbers — is also consistent. On the other hand, it offers, to the mathematician's mind, a sense of familiarity, a possibility of manipulation as an effect of the visual structure of this type of model, and this still further increased the intuitive readiness to accept the rationality of the complex numbers. The geometrical representation of the complex numbers has provided the possibility of using them in physical soiences.

Let us, briefly, mention the basic description of the model.

A complex number $z = x + yi$ is represented by a point in the plane the coordinates of which are x, y. The points on the n-axis correspond to the complex numbers $z = x + 0i$, that is to say they represent, in fact, real numbers. The points on the y-axis correspond to pure imaginary numbers ($z = 0 + yi$). This way, both the real and the imaginary components of a complex number get realistic intuitive interpretations.

The complex number becomes a point and the point in the intuitive code of geometry has a pictorial representation, a spot. And then, this somewhat strange analogy may go on. One may consider the distance between the points and the origin, and then one has:

$$\rho^2 = x^2 + y^2 =$$

$$(x + yi)(x - yi) = z \cdot \bar{z}.$$

The real number $\rho = \sqrt{x^2 + y^2}$ is called the modulus of z and is represented by $\rho|z|$. One then obtains, through the mediation of the geometric functional model, a functional relationship between a complex number and a real number. This is an exceptionally interesting fact. These signals, which seemed to come from a world of ghosts, may be consistently projected in the representable, intuitively acceptable domain of real numbers. These ghosts may be added and multiplied, subtracted and divided.

Adding two complex numbers by analogy with two real numbers one gets: —

$$z_1 + z_2 = (x_1 + x_2) + (y_1 + y_2)i.$$

May this addition be interpreted in geometrical terms?

The answer is: Yes. The point $z_1 + z_2$ is represented by the vertex of a

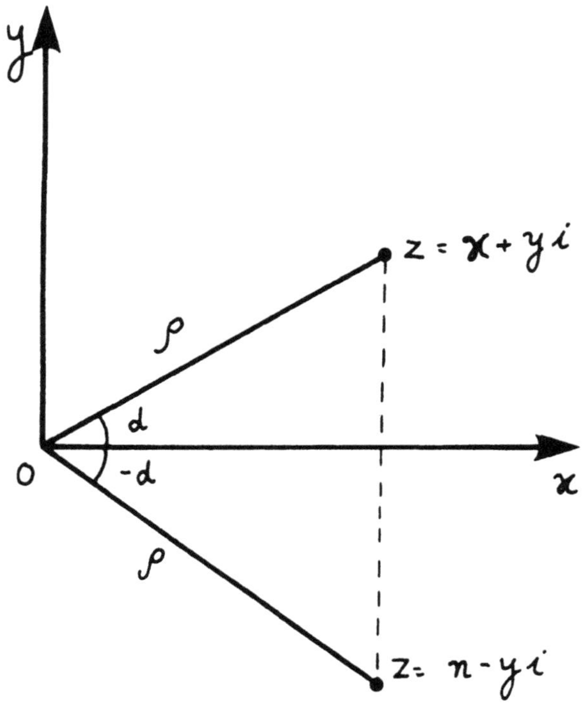

Fig. 6.

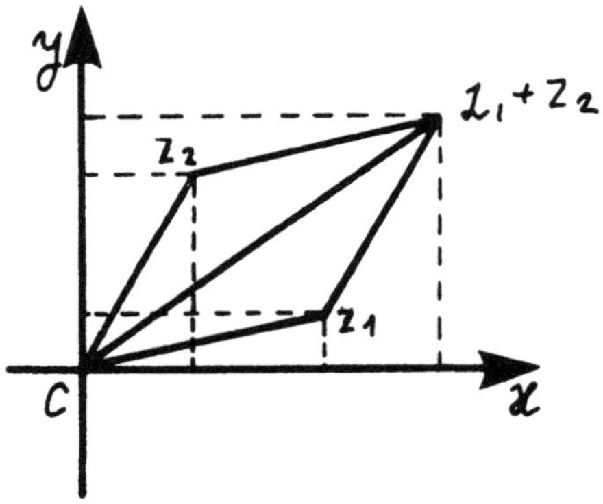

Fig. 7.

parallelogram as shown in the figure. Considering the above figure one may observe that the addition of two complex numbers becomes in representable, real terms, the addition of two vectors like, for instance, the addition of two forces!

One knows that the fact of expanding the concept of number so as to include the "imaginary" numbers, that is to say the creation of the concept of complex numbers, has had enormous consequences for mathematics. It has also played a remarkable role in various technical areas. We do not intend to enter into details, our aim was only to exemplify the profound implications of the intra-mathematical modelling procedures for mathematical reasoning.

The problem with the intra-mathematical intuitive (geometrical) analogies is that one can never be absolutely sure that these types of models do not smuggle into the reasoning process uncontrolled components of the corresponding material-pictorial representations. Let me give more details about the effects of such extra-mathematical models.

Extra-Mathematical Analogies

A third category of analogical models in mathematics is that in which extra-mathematical means intervene. We have already mentioned some examples. Let me quote one more.

I refer to the interpretation of the equals sign. Caroline Kieran (1981) has shown that children tend to confer on the equals sign not the formal, relational properties of equality but rather the behavioral interpretation of an input—output process. By performing a certain activity (the left side) one gets a certain effect (the right side). Various situations follow from this metaphor. For instance, while a child easily accepts a statement like $3 + 7 = 10$, he may consider that the statement $10 = 3 + 7$ is meaningless. An input—output process is not reversible and consequently no symmetry is possible here. Based on the same analogical model, a statement like $3 + 7 = 6 + 4$ cannot be directly accepted as expressing an equality. By adding (behaviorally composing) two sets one does not get other two sets. Children will then interpret the above statement in the following way:

$$7 + 3 = 10; 6 + 4 = 10.$$

In the statement $7 + 3 = 6 + 4$ the equals sign has not for them a meaning by itself (because the input—output model does not apply).

In a very interesting paper James Kaput (1978) has discussed the problem in a broader context.

Let us consider the equations:

$$x^2 + 5x + 6 = (x + 2)(x + 3)$$

$$(x + 2)(x + 3) = (x^2 + 5x + 6)$$

Formally, the two equations represent the same statement because, formally, equality as a relation category implies symmetry. But cognitively the two parts are not symmetrical. We are surprised that children refuse to accept the statement $10 = 7 + 3$ as meaningful and we try to teach them that equality means symmetry. But in the case of the above equation we face a similar problem.

It is not the same to factorize $x^2 + 5x + 6$ with the result $(x + 2)(x + 3)$ and to multiply $(x + 2)(x + 3)$ and to get $x^2 + 5x + 6$. As a matter of fact we are forced to accept the symmetry in the above equation because of formal constraints. Intuitively, we still remain faithful to the process-product metaphor (as Kaput calls it). We still naturally tend to choose one side as the input segment and the other as the output segment. Certainly, if one has to prove the equality one has to start with one side or one has to transform both parts in order to reach identical expressions. What I mean is that, independently of the practical solution, one tends to preserve the tacit, directional input—output interpretation of the equality sign beyond, but together with the formal relational interpretation.

For example (Kaput, 1979, p. 291):

$$\frac{2}{x+3} + \frac{5}{x-2} + \frac{3}{x^2+1} = \frac{7x^3 + 14x^2 + 10x - 7}{x^4 + x^3 - 5x^2 + x - 6}.$$

Brief consideration will show us that, as a matter of fact, subjectively the equals sign here indicates a transformation process more adequately representable by an arrow rather than by an ideal, symmetrical relation.

Mathematicians are not troubled in this case, because they are used to the relational meaning of the equality sign and the tacit, intuitive interpretation is kept under control.

But children may be confused. "You have $10; mother gives you another $5, you spend $7 for a book. How much money do you have at the end?

A child may write:

$$10 + 5 = 15 - 7 = 8$$

This type of error is very frequent in children. The source of the error is the same analogical process-product model. One gets a result and one operates on this result. One does not have to bother about symmetry!

ANALOGIES AS SOURCES OF MISCONCEPTIONS IN MATHEMATICS

We intend now to return to the problem of continuous functions without derivatives. It was Riemann who first found, in 1854, an example of such a function. Weierstrass extended that discovery. In 1872 he mentioned the expression $\cos x + b \cos ax + b^2 \cos a^2x \cdots + \cdots$ in which a is an odd number and b is a positive number less than 1. He showed that its sum is a continuous function for every x and despite this it has no derivative for any value of x if $2ab$ is greater than $2\pi + 2$.

We have suggested earlier that one tends intuitively to believe that a continuous function must have a derivative at each of its points because of the tacit, pictorial model we attach to the corresponding graph.

Let us consider an example taken from a paper of Hans Hahn.
One starts from the figure shown here (8.1).

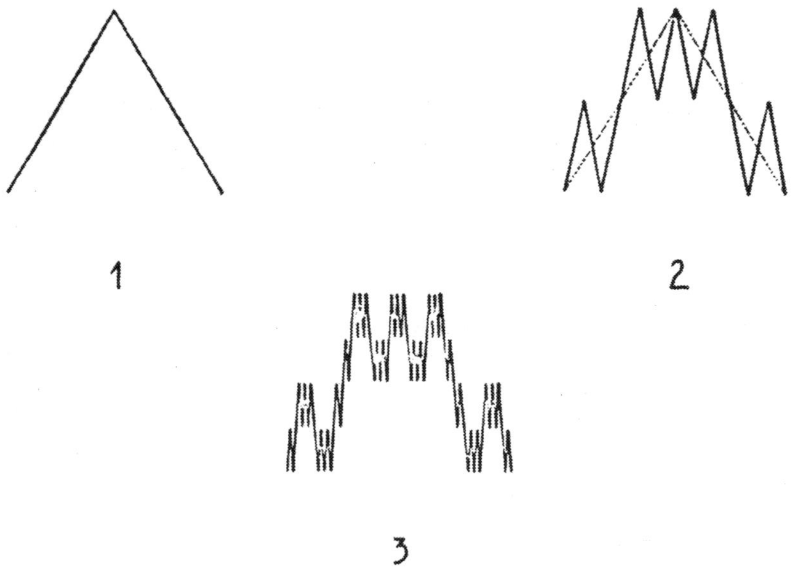

Fig. 8. (After Hahn, 1956.)

One replaces the ascending and the descending lines by broken lines as shown in figure 8.2. One continues in the same manner, reproducing the segments by broken lines composed of smaller segments (figure 8.3) and one continues in the same way indefinitely (Hahn, 1956, p. 1963).

It has been proved that, as the number of segments increases, one approaches a definite curve which will not have a precise slope at any of its points, and therefore no definite tangent. We are mentally able to follow the process of multiplication of segments, we may be ready to accept the idea that the process may go on indefinitely, but we cannot intuitively realize the limit of the process, i.e. the curve itself, because, materially, such a curve does not exist. A curve still remains a fine ink trace in our imagination. A curve without derivatives is not composed of small ink spots — no matter how small — but rather of pure, conceptual, non-dimensional points. Usually, we are not obliged to give up the pictorial representation of a mathematical curve. The analogy may be consistently used for many functions and for many of their properties. Terms like “decreasing” and “increasing”, “minimum and maximum”, “intersection” of curves etc. may be mapped consistently onto a pictorial system of representation. This is the reason why we are not commonly aware that the concepts used are, in fact, supported by pictorial, analogical models.

But there are situations in which the models cannot be permitted to interfere if one wants to obtain a consistent mathematical judgement. Such an effort to discard the mixture of pictorial analogies very often determines conflicting situations which are difficult to resolve. Developments in computer graphics are making it possible to extend the scope of pictorial analogies in fundamentally new ways (see Kaput, in press). For example, facilities can be made available such as "zooming" i.e. being able, flexibly and dynamically, to alter the scale of any particular region of a graph. In the software devised by David Tall (1985), this offers a pictorial representation for differentiable and non-differentiable functions which brings such examples within the range of intuition.

Before presenting a second example suggested by Hahn, let me mention a dialogue I once had with one of my students during a lecture. We have referred to the statement that the sets of points of two segments of different lengths have the same cardinal. The intuitive proof was presented (figure 9) indicating the one-to-one correspondence between the two sets of points. Each of the points contained in *AB* has one and only one correspondent in *CD*, and *vice versa*.

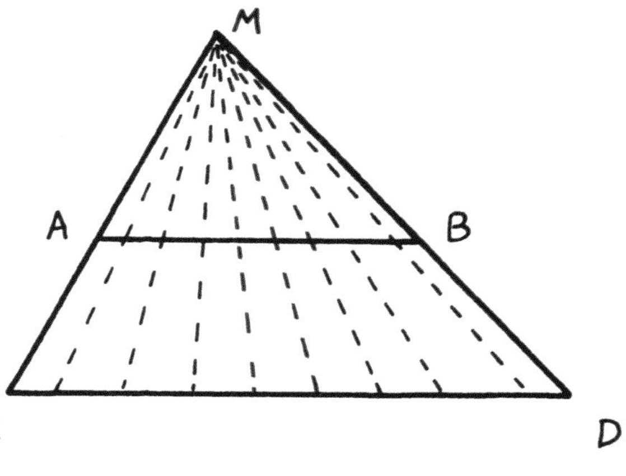

Fig. 9.

Student: You say that the two segments *AB* and *CD* contain the same number of points . . . .

Teacher: Yes, with the meaning that the two sets are equivalent, that is to say, that a one-to-one correspondence may be established between the elements of the two sets.

Student: O.K. You also admit that lines are composed of points and only of points?

Teacher: Certainly.

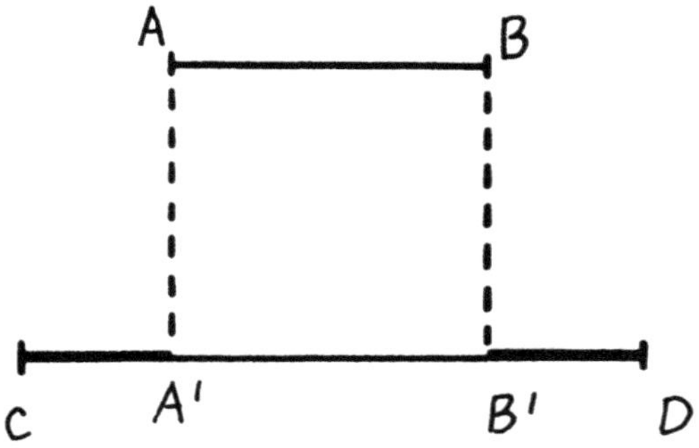

Fig. 10.

Student: Now, if you place *AB* on *CD*, the points contained in the segments *CA'* and *B'D* are not contained in the common part of the two lines.

Teacher: Right.

Student: Where do you find the points for building the segments *CA'* and *B'D*?

Really puzzling! Where do you find these points? There are evidently more points in *CD* than in *AB*!

I try to explain to the student that, while she admits willingly that points and lines have only a conceptual existence, she continues actually to think in figural terms. If points had been real, material spots, then she would certainly be right: the longer lines contain more spots. It is not possible otherwise. But because these points and these lines have no material existence, we must give up the figural constraints, at least under certain conditions, and reason by considering the logical, formal ones only.

My student remained unconvinced. She was sure that she was absolutely fair in considering points and lines only in their abstract sense. But even after eliminating the pictorial connotations, the question "where do you get points for *CA'* and *B'D*?" seems to remain unanswered.

The dramatic, genuinely important aspect of that situation is not the fact that the student does not accept the equivalence of the two sets. What is fundamentally important here is the duplicity of her own reasoning process, about which she is totally unaware.

The reasoning process seems, at the surface, conscious level, to be purified from every interference of pictorial constraints. As a matter of fact, it remains manipulated from "behind the scenes" by the pictorial analogy!

Hans Hahn also refers to the following question: Is it possible to generate entire plane surfaces by a moving point? His guess was that the natural, intuitively acceptable answer would be that a moving point may generate curves but not surfaces. In reality, as he shows, one may prove that moving points, following certain paths, may produce entire surfaces in a finite interval of time. He quotes Giuseppe Peano who, in 1890 showed that such a possibility exists.

Hahn illustrates this claim using some examples, one of which I shall mention here (Hahn, 1956, pp. 1965—1966).

Consider the following three figures (see figure 11).

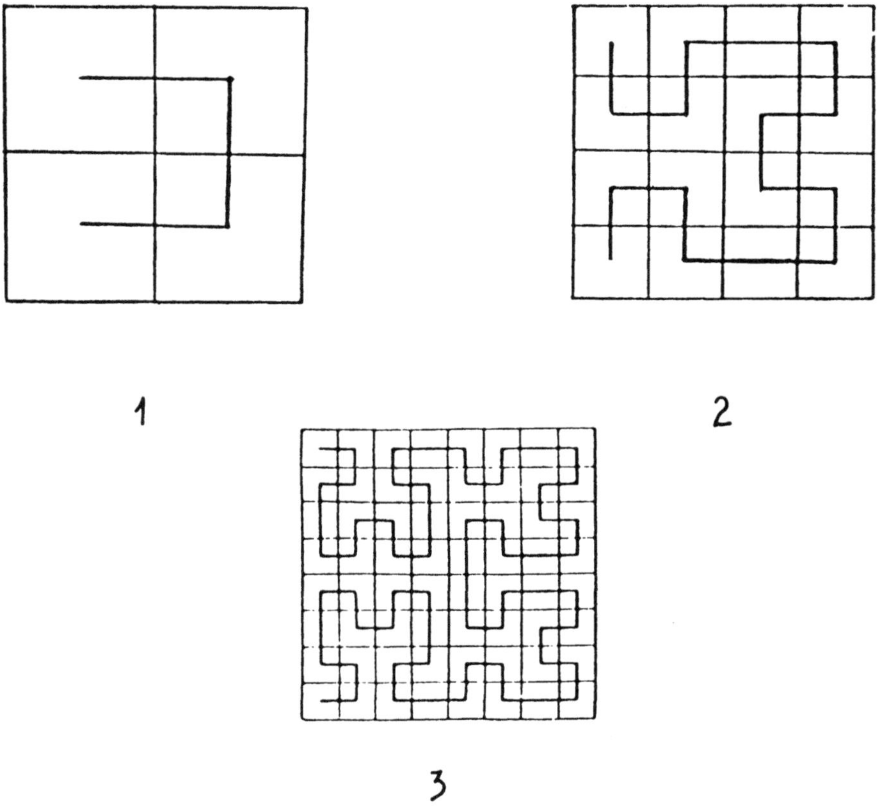

Fig. 11. (After Hahn, 1956.)

The first square is divided into four small squares of equal size. Let us join their center points by a continuous curve composed of line segments; now, imagine a point moving in such a way that at uniform velocity it will traverse the curve in a finite time, say in some particular unit of time. Next, divide each of the small squares into four smaller squares of equal size and connect the center points of these sixteen squares by a similar line and again imagine the point moving etc. One continues in the same way, dividing the squares into smaller ones and joining their centers according to the same pattern. One may prove that the successive motions considered above approach without limit a definite curve that takes the moving point through all the points of the large square in the unit time (Hahn, 1956, p. 1966).

Hahn considers it to be intuitively surprising that a moving point may generate an entire surface in a finite time. In reality, it was Hahn himself who seemed to be surprised by that discovery. Intuitively (if one reasons by genuine intuitive means) it is not surprising.

A fine moving spot, no matter how fine it is, may cover the whole surface, in a certain limited time. The problem appears when one considers the abstract geometrical line, generated by the non-dimensional geometrical point. It has no width and then it is not able to cover even a very narrow strip. The source of the difficulty consists in the fact that, actually, one mixes here two systems of representations: the symbolic, pure mathematical one and the iconic one. In principle, the above problem — the possibility of generating a plane surface by a moving point — should not raise any difficulty if one proceeds consistently either by pure conceptual means or, alternatively, by resorting to a pictorial, material interpretation.

In the first case one formally accepts the equivalence between the set of points of a segment and the set of points of a square. One accepts also that a moving point starting from A may reach B in a limited time.

Fig. 12.

It is then logically acceptable that a point may generate a whole surface in a finite interval of time.

In the second case a small (màterial) spot such as that made by a pencil on a sheet of paper could certainly cover the whole sheet of paper in a finite interval.

The psychological difficulty appears since one tends automatically, almost irresistibly, to assign a kind of material composability to points and segments, while considering them by definition as being pure conceptual entities. But such a composability is meaningful only in practical, pictorial terms.

In fact, the question itself, as it is formulated, suggests a contradictory solution because it is put in contradictory terms.

A mathematical point, having no material existence, cannot move, and certainly cannot connect itself with another immaterial point. By assigning to geometrical entities properties and capabilities which are, in fact, metaphorically borrowed from the real world, we finally arrive, under certain circumstances, at intuitive contradictions.

The psychological impossibility of thinking in pure conceptual terms makes these contradictions unavoidable. Overcoming them means understanding their origin and controlling, as far as possible, their impact on the reasoning process. This is both an epistemological and a very complicated didactical problem.

SUMMARY

Analogies are a rich source of models. To the extent that a structural mapping exists and is known about between the system being studied and another system, analogies contribute in the following ways:

(a) As a heuristic means of generating hypotheses.
(b) By providing an intuitively accessible structure, mental operations which would be difficult in relation to the original system are facilitated. The products of these mental operations can then be interpreted and evaluated in terms of the original system.

Analogies in mathematics may be:

(a) At the formal level, for example, when operations within a domain are analogically extended to a larger domain. For example, polynomials of infinite degree may be treated by analogy with polynomials of finite degree. (Such extrapolations may be fallacious, or may be logically justified much later than they are intuitively accepted.)
(b) Between a symbolic representation and a more intuitive (often geometrical) representation. A clear example of this is the development of graphical representation for complex numbers and the operations on them.
(c) Between a mathematical structure and an extra-mathematical embodiment isomorphic to it, such as the concrete materials devised by Cuisenaire and Dienes.

The other side of the coin is that analogies may be the source of misconceptions when correspondences are assumed which in fact are not part of the structural mapping between the two systems. Often such misconceptions will arise through an incompatibility between a formal property of the system being modelled and an intuitive property of the modelling representation, which is consciously or tacitly guiding the cognitive processes.

CHAPTER 13

PARADIGMATIC MODELS

The notion of representativeness was discussed in Chapter 9 in terms of its contribution as a heuristic to the immediacy of intuitive cognitions. This notion is related to a more general phenomenon which I would term "the paradigmatic nature of intuitive judgment".

If you want to learn the conjugation of the first type of French verbs, you do not learn by hearing "-e", "-es", "-e" etc. You take a representative verb, for instance "chanter", and you learn the forms taken by "chanter" for the different persons. Again, when learning about polygons you refer to *some* polygons which seem to be less biased by certain particular features. You choose, for instance, a pentagon or a hexagon with unequal sides and angles for studying the general properties of polygons, and not squares or equilateral triangles.

A paradigm is not a mere example; it is defined by its function rather than by its intrinsic qualities. It is a particular instance or a sub-class of a class of objects, used as a model. However, a model is not necessarily a paradigm. For example, a tree diagram used for solving combinational problems is a model, but it is not a paradigm. Moreover, not every example may play the role of a paradigm. A pupil asked to exemplify a certain concept (let us say the concept of mammals) is not asked to produce a model of the concept but to prove that he understands the term and uses the corresponding concept correctly.

On the other hand, if one has to define a concept, that is to say to describe its general properties, such as when one has to use the concept in relation to others in a productive reasoning process, this concept never works as a pure logical construct. The meaning subjectively attributed to it, its potential associations, implications and various usages are tacitly inspired and manipulated by some particular *exemplar*, accepted as a representative for the whole class. There are such particular instances which confer their genuine productive capacity on concepts. A definition or a formal description is never sufficient for really understanding the meaning of a certain term. One must always ask for examples. Through the paradigm, the concept gets, subjectively, a structured meaning; the various properties defining a concept do not appear to be arbitrarily related. They describe a certain object and it is this object which, being kept in mind, confers unity, structure and productiveness to the concept. Paradigms play a fundamental role in every intellectual activity: in defining, in understanding, in learning and in every solving attempt.

Experimental work on concepts has shown that the "classical view" of concepts as definable by a set of features which are singly necessary and jointly sufficient, is impossible to reconcile with the ways in which people actually process concepts (see Smith and Medin (1981) for a very useful summary). Most recent theoretical attempts to supersede the classical view have been based on prototypes (notably in the work of Rosch, e.g. Rosch (1978) or on "the exemplar view" which holds that concepts "are represented by their exemplars (at least in part) rather than by an abstract summary" (Smith and Medin, *op. cit.*, p. 143).

In all these situations, the particular instances usually attached to a concept by a person, play an active role in shaping meanings, interpretations and connotations. They inspire assumptions, rich strategies, conclusive decisions. Their tacit impact on the reasoning process is much more important than is generally assumed. When thinking about thinking, you have a particular type of thinking in mind — for instance the solving of problems. When one tries to define what solving a problem means, what the main steps are in solving a problem, one considers, in fact, a particular type of problem as being representative; for instance, chess or arithmetical problems. This is trivial, but what happens is that *we tend to see the whole class, the entire concept, through these special instances.* When one considers the concept of a model, it is a certain type of model which is manipulating our definition and interpretative attempts, for instance analogical models. Moreover, it is usually a particular exemplar, subjectively the most familiar, which tacitly becomes "the model" — for instance, the analogy between the structure of the atom and the planetary system.

When people are arguing about the venality of physicians in the modern world, they do not generally have in mind objective, statistical data but only a few, familiar examples. When you try to describe "the English" in comparison, for instance, with "the French" — and even to draw some political conclusions — what you usually have in mind are mainly examples of *certain* English and French people you have met or with whom you are acquainted. Alternatively, one may think in terms of a national stereotype (or prototype) such as "John Bull". Despite this, one tends to generalize and to believe spontaneously in the general validity of the conclusions drawn.

Paradigmatic tacit models are, then, an essential factor in shaping our intuitive approaches, interpretations and solutions. There is a deep analogy between the role of paradigms — as defined by Thomas Kuhn — in scientific revolutions, and the role of paradigmatic models in an individual's productive reasoning activity. I would tentatively affirm that important turning points in our ideas are, generally, related to the discovery of certain *particular* phenomena which throw new light for us on the significance of a whole category of data.

Hershkowitz and Vinner (1982) have reported a number of findings which represent a good illustration of the role of paradigmatic models in the mathematical reasoning of students. Indeed, I would say that their findings constitute good paradigms for what a paradigmatic model means.

A group of 189 students (grades 6 to 8, age 11—14) were asked to construct an altitude to the side '*A*' in a number of triangles (see figure 13).

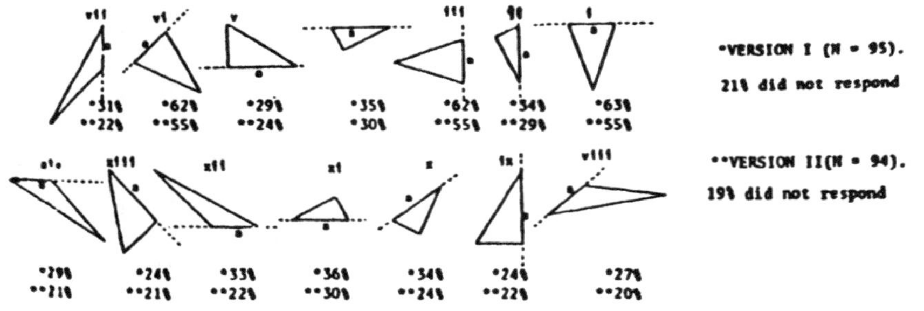

Fig. 13. (After Hershkowitz and Vinner, 1982.)

The numbers adjacent to the various triangles indicate percentages of right answers. There were two versions. In the first version the following definition was given: "An altitude in a triangle is a line from one vertex perpendicular to the side opposite to this vertex or to the extension of this side". In the second version no definition was given. The two percentages quoted in each case refer to the first and second versions respectively.

As can be seen, the success depends on the type of triangle. For instance, in the case of an isosceles triangle the correct answer was given by about 60% of the students, while for a right triangle only about 24% answered correctly. It can be seen that if the altitude falls outside the triangle this has a negative effect (see for instance, items 7, 14, 12). It can be seen also that the presence of the definition produces an increase in the percentage of correct answers, but the differences are not very great (about 20%) (Hershkovitz and Vinner, 1982, pp. 20—23). A clear hierarchy may be deduced from the following table which refers to the same task (figure 14).

The main conclsion of this line of research is that subjects attach a certain, particular, representation to the concept of altitude in a triangle which seems to have a strong impact on their cognitive decisions. *Even when the definition was explicitly mentioned*, most of the subjects were still not able to respond correctly. It was also found by Hershkovitz and Vinner that about 20% of the students drew the altitude as a median in items 2, 3 and 4 (see Fig. 14). One may deduce from this finding that, to many subjects, the paradigm for the concept of altitude is specifically related to isosceles triangles. Even in the case of item 2, where the altitude from vertex *A* is inside the triangle, there are only 40% correct answers. Of those giving incorrect answers, 20% drew a median, about 20% gave various incorrect answers, and 20% did not

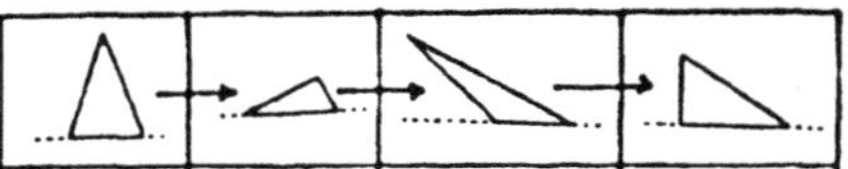

no construction	5.5%	8%	7%	11.5%
altitude to *a*	74%	40%	32%	30.5%
median to *a*	-	20%	21%	20%
perpendicular bisector to *a*	-	7%	7%	9%
altitude but not to *a*	-	1.5%	6%	4%
the side *a* itself	3%	2.5%	3%	3%
others	17.5%	21%	24%	22%

Fig. 14. (After Hershkowitz and Vinner, 1982.)

answer at all. One may suppose that the distribution of the various categories of solutions may be influenced by age and instruction. Our intention was only to stress the fact that *the manner in which a concept functions in a reasoning process is highly dependent on its paradigmatic connections. The fact of knowing explicitly the definition does not eliminate the constraints imposed by the tacitly intervening paradigm.* Let me cite an example to illustrate this. "Given triangle *ABC* and the median *AD* prove that the areas of the triangles *ABD* and *ADC* are equivalent." (fig. 15) A frequent, initial, tendency is to try to determine the two areas by considering the altitudes drawn from the vertex *A* for triangle *ABD* and vertex *D* for triangle *ADC*.

In fact, there is a simple direct solution. In both triangles, the bases are equal ($BD = DC$) and there is the same altitude drawn from *A*. The difficulty with the *ADC* triangle is that the altitude drawn from vertex *A*

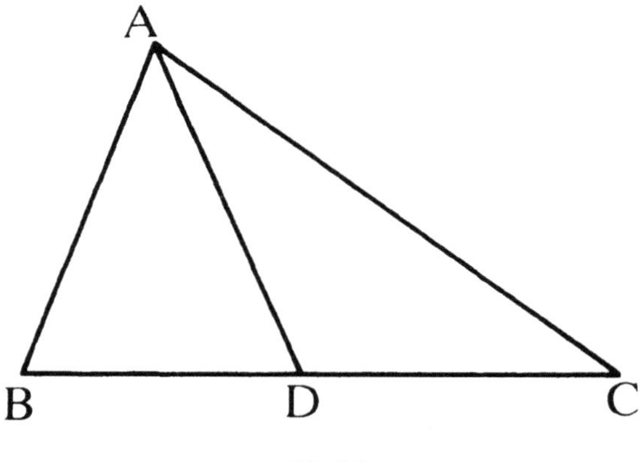

Fig. 15.

strongly deviates from the usual paradigm. It follows that the paradigm attached to a concept not only affects the way in which particular instances are identified, but may also affect the strategy adopted for solving a certain problem and, very often, the solution itself.

An important effect which paradigmatic instances may have on conceptual structures is a phenomenon which may be called the *dissolution of hierarchies.*

Students are loath to agree that numbers like 2 and 3 are complex numbers. They would identify them as natural numbers or rational numbers despite the fact that they would agree that $x + iy$ is the general form of a complex number and that y may be equal to 0 (See Tall and Vinner, 1981, p. 154).

To many pupils the notions of parallelogram, rectangle and square are not hierarchically organized. They represent classes of quadrilaterals with *the same level of generality.* This is because the images usually attached to each of these notions act not as particular instances but as general models. The paradigmatic model of parallelograms is usually that which appears in Figure 16 in which unequal adjacent angles are implicitly stated as necessary properties of parallelograms.

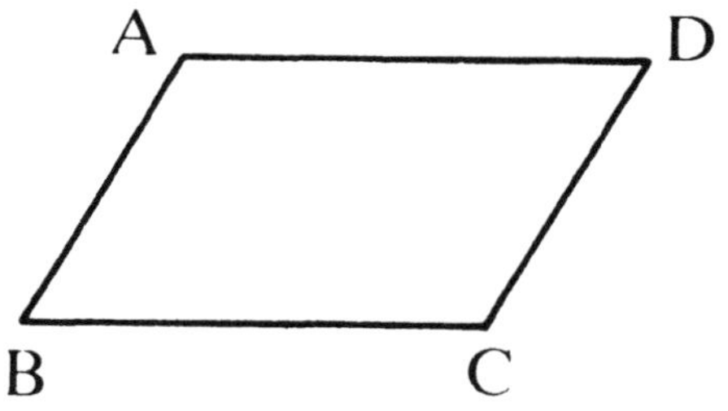

Fig. 16.

The next example reveals the interesting, conflictive situation which may appear as an effect of contradictions between the intuitive and the conceptual level. Shlomo Vinner presented students with the following three curves (see Figure 17) and asked them to answer whether it is possible to draw one, two, or more tangents to these curves through the points indicated on the drawings. They were also asked to draw these tangents and to define the concept of a tangent (Vinner, 1982).

Figure 17 shows the typical solutions given to the different items.

The subjects were 278 first-year college students enrolled in calculus courses in Chemistry, Biology, Earth Sciences and Statistics. As Vinner has pointed out in his paper, most of the students seem to have in mind a particular representation of a tangent, *namely a tangent to a circle.*

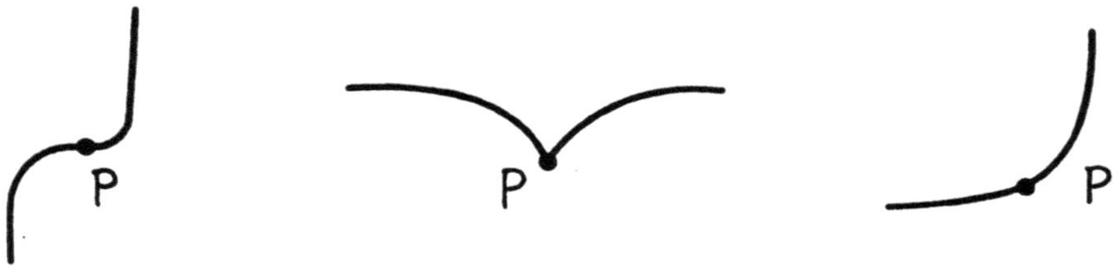

Fig. 17. (After Vinner, 1982.)

Distribution of student drawings
(N = 278)

The right answer		2 tangents	Another drawing	No drawing
18%	38%	6%	10%	28%

Table 3 - Distribution of student drawings to question 2
(N = 278)

The right answer	2 tangents	Infinitely many tangents		No drawing
8%	18%	18%	14%	42%

Table 4 - Distribution of student drawings to question 3
(N = 278)

The right answer		2 tangents	Infinitely many tangents	Another drawing	No drawing
12%	33%	16%	7%	4%	27%

Fig. 18. (After Vinner, 1982.)

This may be inferred from the fact that, in most situations, the students avoided drawing the tangent in contact with what seems to be a straight segment of the curve (items 1 and 3). Secondly, students consider tangents to be *lines* which apparently touch only one point of a curve, independently of the fact that these lines have nothing to do with the slope of the curve at the given point.

The percentages of correct answers were 18% for curve 1, 8% for curve 2, and 12% for curve 3. At the same time 41% of the subjects offered correct definitions, as taught in their calculus courses. This means that for many of these 41%, the real meaning of the concept of a tangent was determined not by the apparently known definition but by a particular case (the tangent to a circle) used as a paradigm.

The conflictive nature of the situation (between the paradigm and the concept) was revealed in various types of reactions.

Item 1: 6% of the subjects drew two lines, one correct and one responding to the "circle paradigm" constraint.

Item 2: 18% of the subjects drew two lines both through point P, apparently, attempting to reflect the different slope of the two branches of the curve.

Item 3: 16% of the students drew two lines, one correct and one fitting the circle paradigm demands.

It is not only concepts that are paradigmatically loaded. The same may very often be said about statements and reasonings.

What do you mean by: "$p \rightarrow q$" (p implies q)? I mean by that statement that from the truth of p necessarily follows the truth of q. But what about a situation in which q is denied? What about p then? One may be perplexed by such a question (if one is not used to logical exercises) first of all because the term "implies" is loaded with a behavioral meaning — p is the agent, the cause, and q is the effect. By reversing the roles one gets a question with no intuitive meaning. In order to be able to answer one needs an example. For instance "If object A is a metal, then it is electrically conductive. By checking object A I have found that it is not electrically conductive. Therefore it is not a metal."

One obtains a solution *to the logical problem*: "given $p \rightarrow q$, what about p if q is defined?" by resorting to an example. But this example is, in this case, a model. One solves the problem in terms of the model and one gets a solution in terms of the original (the logical question). The above particular concrete example, or another, may become the usual, practical content to which I may always refer when having to remember the truth table of implication in order to check whether a certain practical relation is an implication or when solving a problem which contains an implication.

It is very well known that concepts and formal statements are very often associated, in a person's mind, with some particular instances. What is usually neglected is the fact that such particular instances may become, for that person, universal representatives of the respective concepts and statements and then acquire the heuristic attributes of models.

Two people who argue about a certain problem may sometimes disagree because the notions they use are based on different paradigms.

We have used the term paradigm in the present text in a way which is very similar to that in which Thomas Kuhn uses it. "By paradigm", he writes, ". . . I mean to suggest that some accepted examples of actual scientific practice — examples which include law, theory, application and instrumentation together — provide models from which spring particular, coherent traditions of scientific research" (Kuhn, 1970, p. 10).

The difference is that Kuhn uses the term "paradigm" with reference to sociological-historical frames, while in the present essay the term paradigm refers mainly to a personal, subjective experience. In both cases, as in the original linguistic meaning, paradigm means the usage of examples as *exemplars*, as models for supporting the processing of universals.

It seems to me to be of very great psychological interest that, at both levels — the individual and the sociological—historical one — *the process of reasoning, at least in its genuinely productive forms is, to a very great extent, stimulated, shaped, and controlled by paradigmatic instances*. One may plausibly affirm that the fundamental role played by paradigms in scientific revolutions has its roots in this fundamental psychological phenomenon: one thinks about universals in terms of specific, structured instances. We are generally not aware of this double game because we usually see the universal through the particular. This is one of the main features of intuitions.

Kuhn says that the scientific paradigm includes law, theory, application and instrumentation. It is the same with current individual paradigms. The particular instance of a tangent to a circle is, for many students, the tangent. It becomes the tacit theory of what tangents are: lines which touch a curve at only one point. The essential property, that it expresses the slope of the curve at a certain point, is forgotten. That intuitive, implicit theory of the student is related to a certain "application" and "instrumentation". As long as you are not able to draw a line which touches the circle at a single, given point, you do not in fact understand what a tangent means (in its primitive meaning). A paradigm is never a simple image. In order to effectively exert its role as a source of intuitive understanding, the paradigm must synthesize an iconic structure with procedural prescriptions.

Let us now consider a different case. Let us suppose that a subject is asked to determine the class to which a certain group of objects belongs. It is important to establish whether the subject's answer (object a is an A) expresses the fact that to him A is truly a class containing a or only that A is a *paradigmatic model*. The difference is fundamental for understanding the student's conception in the respective domain.

Let me come back to an already quoted example. In a recent study carried out by Stavy and Stachel, children aged from 5 to 15 were asked, among other problems, to say if a melted candle is inflammable. "Would the melted substance burn if lit?" Many of the children who affirmed that the substance would not burn justified their answer by claiming that, after melting, what one obtains is water.

Stavy and Stachel are inclined to believe that for many of the children who offered that explanation the melted candle is really water and therefore they see an analogy between this interpretation and the theory of Thales according to which water is the fundamental element from which reality is built (Stavy and Stachel, 1985).

My hypothesis is different. I think that for most of the children who identified the melted candle with water, water represents only a *paradigmatic model for liquids*. They do not possess the term *liquid* though they may have the concept. They use the term water because water is the most familiar and the most representative liquid substance. I have no experimental evidence for that specific case. But it is important, as I said, to determine clearly what the child has in mind when using the term *water* in the above experiment. Didactical referring to the conservation of substance, of weight and of other properties, after changing the state of matter, cannot be correctly solved without clarifying the real meaning for the child of his answer. Our point of view is that paradigmatic models may play an important role in children's definitions, explanations, classifications and predictions and therefore they have to be identified as such. Sometimes the paradigmatic model may genuinely influence the definition given by the subject for the entire class but sometimes it may be only "une façon de parler".

Mason and Pimm express the following ideas with regard to the relation between the general and the particular in the pupils mind:

A teacher having written an example of a technique or theory on the board is seeing the generality embodied in the example, and may well never think of indicating the scope of the example, nor of stressing the parts that need to be stressed in order to appreciate the exampleness. However, the pupils have far less experience even with a particular instance of the situation under discussion (and may well be unaware there are others) which as a consequence absorb all their attention. The pupils may see only the particular (which is possibly for them still quite general, i.e. not mastered). As a result they often try to 'learn the example'." (Mason and Pimm, 1984, p. 286.)

What Mason and Pimm do not explicitly say is that paradigmatic representations of concepts are usually unavoidable not only in students but in experts as well. Paradigms are a part of our productive thinking. The problem is not that sometimes the student does not see the generality through them. They are a part of our intuitive mechanisms and in fact they do carry general meanings. The problem is that the tacit boundaries of the concept suggested by the paradigm may differ from person to person, depending on their personal experience and their information, and that the paradigmatic definition of concepts is mostly an uncontrolled process.

A pupil who refuses to accept that figure *ABCD* is a parallelogram, does

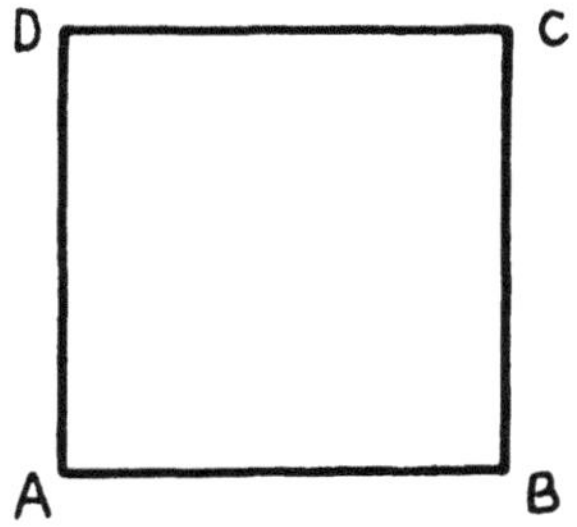

Fig. 19.

so because, to him, the inequality of angles and sides in a parallelogram are general properties related to this concept. Therefore, the problem with a paradigmatic model is not the capacity to grasp a general meaning (this happens automatically) but to identify those properties which are qualified as general in the realm of certain conventions (for instance, scientific conventions). The psychological conflict is not between a general and a particular meaning but between two (or more) general meanings which might be suggested by the paradigm.

The properties tacitly selected as general ones certainly depend on age — in addition to experience and instruction — and perhaps, on some personal characteristics. It has been shown, for example, that to young children, pictorially striking features may be decisive, while older children may be more influenced by relational properties. (Olver and Hornsby, 1966).

We do not possess experimental evidence concerning didactical means which can overcome the difficulties engendered by the influence of paradigms. The following ideas may be envisaged. Firstly, that a child has to learn early enough what a definition means — starting as early as during the concrete operational period and certainly fixing and experiencing the concept of definition during the formal operational period. The child has to learn the decisive role of *explicitly defining concepts* as a sine-qua-non way of avoiding errors in using the corresponding terms. My belief is that a twelve year old child may understand the dual significance of *meaning*: On the one hand *the definition* explicitly stated (which are the properties defining a class of objects?). On the other hand the *exemplar* (or the exemplars) able to confer structurality and intrinsic coherence to the concept. By analyzing the paradigm in the light of the concept, by learning to find correct examples and counter-examples corresponding to the concept, the student may reach this stage of grasping a concept which is not void, related to exemplars which are not misused.

This training activity must become, in my view, a part of every discipline taught in school. It contains a general training, leading the student to the awareness of the fact that concepts are very often defined on the basis of certain paradigms and that these paradigms may determine tacit generalizations which are not in accordance with formally accepted conventions.

Secondly, the training must be specific as well. The student has to learn *practically* to identify the paradigms to which his conceptions are related, and to check the validity of the generalizations made, by confronting his personal representations attached to a concept with the properties stated by the definition.

When you affirm that a *parallelogram is a quadrilateral* the opposite sides of which are parallel (or with opposite equal angles) this is *exactly* what is meant by a parallelogram. Nothing is said about adjacent angles or sides. They may be equal or not. It, then, follows that such an analysis implies both aspects, identifying the general properties and making explicit what is not implied by the concept but may be accidentally suggested by the paradigm used.

I think that such a training, starting from childhood, may be of considerable importance not only for the intellectual development of the students, but also for their social and ethical education.

CHAPTER 14

DIAGRAMMATIC MODELS

A third category of models to be mentioned is that of diagrams. Generally speaking, diagrams are graphical representations of phenomena and relationships amongst them. Venn diagrams, tree diagrams, and histograms used for statistical representations, belong to that category. While analogies usually represent mappings between two existing, relatively independent systems, in the case of diagrams one system, the original, exists in its own right while the other, the diagram, is an artificial construct, intentionally created to model the first.

A diagram possesses important intuitive features. Firstly, it offers a synoptic, global representation of a structure or a process and this contributes to the globality and the immediacy of understanding.

Secondly, a diagram is an ideal tool for bridging between a conceptual interpretation and the practical expression of a certain reality. A diagram is a synthesis between these two apparently opposed types of representation — the symbolic and the iconic.

Consider Venn diagrams and the ways in which they are used.

For instance, relations like inclusion and equality, operations like union and intersection, and their respective properties, receive visual representations which directly suggest, those relations, operations and properties.

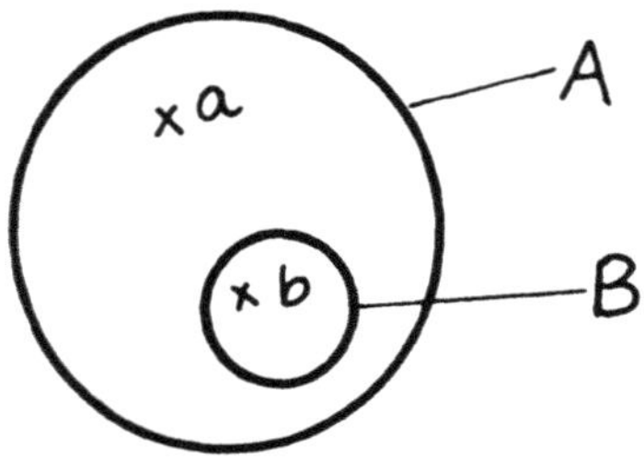

Fig. 20.

In order to be able to use the diagrams fruitfully one has to establish a number of conventions which specify the meanings of the images used.

Consider two sets, A and B, with B being included in $A(B \subset A)$ and a and b generic elements of A and B, respectively. One has explicitly to define the notion of inclusion by stating that a set A includes a set B if every element of B belongs to A as well. This definition does not result from merely inspecting the image. One has to understand that the elements belonging to B possess *all* the properties of A and possibly some other additional specific properties. All squares are rectangle but not all rectangle are squares.

Things are not so simple as they may appear at first glance. The terms *inclusion, union, belonging* to, etc., are used with metaphorical meanings. They convey what were originally notions with *practical, behavioral* meanings. These get logical *interpretations* only through a system of explicit conventions.

The class of rectangles includes a subclass, that of squares. A house may include a living-room and the price of an object may include sales tax or value-added tax (VAT), or not. In these examples the term inclusion has different meanings. Sets do not include subsets as a material part of them, but as a logical relationship between classes. The essential difficulty is that in the set terminology a "*b*" is *at the same time* a "*b*" and an "*a*". (The class of elements "*b*" is included in the class of elements "*a*"). A living-room is not at the same time a living-room and a house and does not even have the properties of a house. VAT is not at the same time VAT and the price paid for a certain object. The logical meanings of inclusion, of "to belong to", etc. are different from their practical significance. The same practical significance excludes the possibility of considering two sets *A* and *B* as being at the same time in a relation of inclusion and in a relation of equality. One may formally use the statement "*A* includes *B*" even if *A* does not possess elements which are not in *B* (all the elements of *A* are *B* and *vice versa* — the two sets are equal).

All these specifications do not result directly from inspecting the Venn diagram currently used for inclusion.

Considering the diagram, the learner may assume that *B* is included in *A* as the stone of a plum is included in the plum.

With such an interpretation in mind he may assume that *A* is composed of two different reciprocally exclusive structures (like the edible part of the plum and the stone). Certainly he will not assume spontaneously that the above figure may suggest that sometimes the two components *A* and *B* may be identical. He is directly led to assume that *A* is necessarily bigger than *B*.

In other words, the use of Venn diagrams, if unsufficiently prepared, may complicate the active understanding of set concepts and relationships instead of rendering them more intuitive.

We mentioned in Chapter 13 that the paradigmatic structure of a productive reasoning process (the natural use of examples as models) may bring about the dissolution of conceptual hierarchies. It is clear that the use of Venn diagrams may help a better understanding of such hierarchies (and the operations with them) only if this diagrammatic tool is well assimilated on the basis of explicit convention. Very often that is not the case. Most pupils still have difficulties in understanding that a square is at the same time a rectangle *and* a rhombus and that a square, a rectangle and a rhombus are *all* parallelograms, etc.

If the conventions are not perfectly understood, the pupil will have great difficulties using the Venn diagrams for various modelling purposes, even for very simple instances. If $B \subset A$ then the union of the two sets is A and the intersection of the two sets is B. Such a simple proposition is not easily understood without a Venn diagram. The diagram makes the two assertions *intuitively clear* but for this one has to understand correctly the language, the figural symbolism used. And as I said, *the symbolism used is not intuitively evident*, first of all because it contradicts what is naturally, practically acceptable.

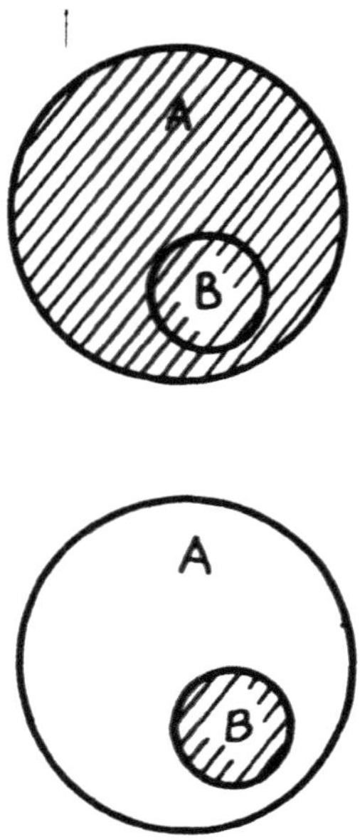

Fig. 21.

Let us think first of $A \cup B$. The notion of a union containing both A and B has a clear, practical meaning if the two sets are mutually exclusive. By putting together 4 red marbles and 3 green marbles one gets 7 colored marbles. But what the above model says is that by joining 7 colored marbles with 3 green marbles one gets 7 colored marbles. This seems mere nonsense from a practical viewpoint.

Similarly with intersection. Firstly the image does not seem to suggest any intersection at all. In terms of direct, figural representation, the word *intersection* suggests the notion of crossing lines. The above lines do not cross. It is only through a formal convention that one may refer to intersection with regard to the above image. What the above diagram says is the following: if you have 7 colored marbles and among them 3 are green, then the common part of the two sets are the three green marbles. One has to make some effort to understand the meaning of this sentence. One finds at the end that it is correct, but what a waste of words to say almost nothing! From a practical point of view the above sentence seems useless and, consequently, meaningless.

This shows that the intuitiveness of the above diagrams is not a natural property. One cannot rely directly on the respective images in order to get a more convincing grasp (in the intuitive sense) of the concepts and relationships expressed. The image is not directly readable in terms of the original.

Diagrams are not, generally, the direct image of a certain reality. If one wants to get an intuitive feeling of what 'speed' means one has to watch a moving body or, better, to compare two moving bodies. If one wants to get an intuitive feeling of how the law of large numbers works in probabilistic situations one effectively has to repeat a stochastic experiment many times and to record series of outcomes.

But with diagrams things are totally different because a diagram, although expressed in figural terms, is not a primary cognitive instance. *It is the figural expression of an already elaborated conceptual structure, like any other symbolic system.*

The main source of the modelling capacity of diagrams is the natural, fundamental isomorphism between the dynamics of concepts and the dynamics of spatial relationships.

One interprets logical disjunction in terms of combined areas, logical conjunction in terms of shared areas. The complement of a set B included in a set A is the part of the A area outside the B area etc. In fact one uses already existing metaphors and one fixes strict principles of conversion.

The spatial image obtained obeys its own figural rules and its own figural logic in a consistent way. A point cannot be inside and outside of a closed curve; a point which belongs to two areas belongs to the shared zone of the two areas; if a point belongs to B and B is included in A then that point belongs to A etc. All these rules are figurally evident. If one is able to translate the content of a figural statement in terms of a figural—diagrammatic representation, one gets *a figural statement which is directly, intuitively acceptable.*

The diagram may work as an *heuristic model*, that is as a device for solving problems, because of its isomorphic correspondence with the original and because of its autonomy in respect to the original. One translates the problem raised by a situation A into terms of diagram B, one solves the problem in B and one reinterprets the solution in terms of A. For example, there is evidence (e.g. Quinton and Fellows, 1975) that some subjects solve problems of the type:

$$A > B, B > C\text{: what is the relationship between } A \text{ and } C?$$

(where $>$ represents some transitive relationship) by constructing an image of a vertical line on which *A* is positioned above *B* and then *B* is positioned above *C*. The relationship between *A* and *C* can then be "read off" from the image and interpreted in terms of the original problem.

Actually, a diagram may directly express certain spatial relationships. Inclusions, intersections, directions, positions etc. are all directly perceptible. But this does not imply that the respective image also delivers automatically and intuitively the message of the original reality which it is intended to reflect. *A diagram is necessarily a post-conceptual structure.* It describes, in figural terms, conceptual relationships which, in turn, are the symbolic processed expression of an original reality.

In order to be able to take advantage of the intuitive virtues of the diagram one has to possess a perfect command of the entire system of the conversion principles.

Diagrams belong to the "symbolic mode" (in Bruner's terminology). *But they possess the exceptional quality of conveying the message in a structured, iconic way and this confers on them a highly intuitive potentiality.*

Let us mention a second example — tree diagrams used for solving combinatorial problems. How many arrangements of 3 elements each can be obtained using two digits, for instance 1 and 2 (with repetition)?

The diagram shows the solution.

There are a few very simple conventions: in column *A* the two possibilities of the first digit (1 or 2); in column *B* again the same two possibilities for the second digit; and in column *C*, again, these two possibilities. The idea is to list for the three positions of a three digit number each of the two possibilities prescribed by the problem. The multiplicative structure of the solution becomes evident (in the above example $2 \times 2 \times 2$).

Certainly, there is no natural analogy between the tree diagram and the various possible arrangements. In its own graphical terms, the tree diagram translates, through a system of conventions, the original process of building sets of digits. The translation takes place through the agency of a conceptual structure. One has to be aware that the same element remains available throughout the whole experience and therefore after each digit in the diagram one has a fan of arrows leading again to each of the originally given elements. The diagram represents a kind of flow chart of the conceptual process of building the sets of digits rather than graphical duplication of already identified groups. The symbols 1, 2 have in fact to be conceived as symbols representing generally two distinct elements irrespective of their nature.

As a matter of fact a pupil has to learn the construction of the diagram and the meaning of the image obtained at each step. He has finally to learn to translate the *tree* obtained with its fans of possibilities into sets of possible arrangements. Our experiments have shown that a 10-year-old child is able to grasp easily the principle of the tree diagram which mediates his understanding of combinatorial procedures. It has been shown that the child is

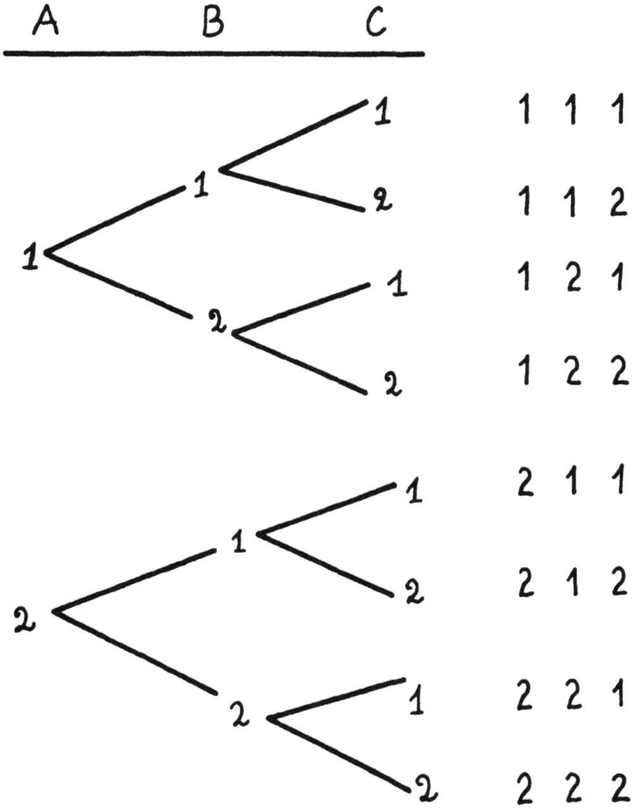

Fig. 22.

able, after realizing the first step, to generalize the procedure and to solve arrangement problems (with repetition) with various numbers of elements (that is A_n^m with repetition in which m and n may take different values). Moreover, the experiments have shown that ten-year-old children are able to generalize, spontaneously, *the principle* of the diagram and to solve permutation problems as well (Fischbein, Pampu and Minzat, 1970).

Let us now pass to a different type of diagram, graphs representing functions. In this case too, as generally in diagrams, the correspondence between the original and the model is not grasped directly as an effect of a natural similarity (as happens in analogies). The model, the graph, is a post-conceptual construct which translates by strictly defined conventions the properties of the original in terms of a figural representation.

If one considers, for instance, the graph representing the relation between time and space in the case of falling bodies there is no direct, sensorial similarity between the phenomenon of falling and the form of the graph. The graph represents, rather, a function (a conceptual structure) representing, in turn, the constant relationship $s = s_0 - \frac{1}{2}gt^2$. No direct interpretation of the graph is possible (in terms of the real phenomenon) without an understanding of the conceptual intervening structure (the mathematical function). Although a graph is a diagram according to our definition (a graphical-conceptual modelling construct) it also may be considered that graphs representing functions are in fact related to the class of analogies (as in Polya's interpretation). In fact, the analogy we are referring to is *not* between the original phenomenon and the graph but rather between *the numerical expression of the respective phenomenon* and its graphical (in fact spatial) representation. The geometrical and the numerical system of entities are *not* governed by mere arbitrarily chosen conventions. On the basis of adequate axioms the world of numbers and the world of figures behave, each of them, in an absolutely coherent, internally consistent manner. Although reciprocally independent, the two systems appear to be at the same time totally isomorphic. The two systems, the numerical and the figural ones represent an ideal analogy, probably the best we know in science. But it is by no means possible to consider the relationship between the original phenomenon and the graph representing its function as an analogy. A graph is a *diagram* which *uses* the analogy between the numerical system and the system of geometrical properties.

This sounds rather complicated and, in fact, it probably appears complicated to the student. A graph with its figural properties very often has the properties of a Gestalt. It imposes itself on the learner as a *figure*, in the Gestalt sense, as a structured directly interpretable reality. For that reason it should represent an excellent intuitive device. In fact, a graph is not, by itself, generally an intuitive device. Like other types of diagrams the graph is neither an example nor an analogy in respect to the phenomenon to be represented. As already said, the relation between the graph and the original is indirect, it takes place through an intervening conceptual structure. A graph may become an intuitive device only *after* the system of conventions relating the original reality, the intervening conceptual system (the function) and the graphical representation have been internalized and automatized.

Consider again the very simple example quoted above of a graph representing the displacement of a body moving with constant acceleration (for instance a falling body). The graph is a parabola. The natural tendency of a novice would be to confuse the form of the curve with the trajectory of the motion. The graph is a "good shape" in Gestalt terminology (although it is not a closed one) so real, so directly interpretable that it is actually difficult to escape its direct intuitive constraints, in order to grasp its indirect message (the constant growth of velocity). But after the system of conventions has been internalized and automatized the image may really help to get an

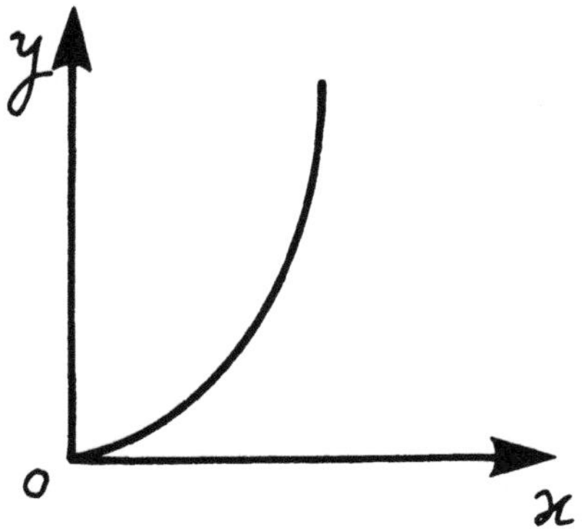

Fig. 23.

intuitive view (synoptic, internally consistent) of how position is changing with time in an accelerated motion.

Let us consider now the graph representing the motion of a body with constant velocity (fig. 24). What the graph — the straight line — represents is the direct proportionality between space and time in this case. As time doubles, space doubles as well. In equal intervals of time the body travels equal distances. What the parabola means, in the case of constantly accelerated motion, is that the displacement is *not* proportional with time and that *during* successive equal segments of time the displacements are constantly increasing.

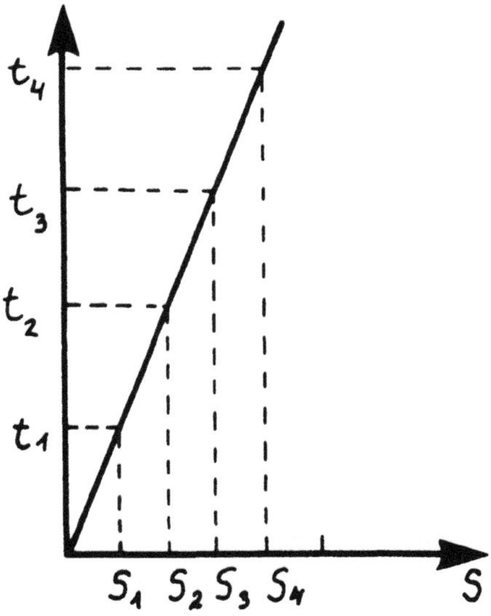

Fig. 24.

Those who are used to the language of graphs may be really helped to obtaining an intuitive view of a phenomenon. On the other hand the intrinsic intuitive properties of a graph may also represent a source of misinterpretations because the graph constitutes a self-sufficient figural system *with no natural appeal for an extrinsic meaning*. With no open valencies, the impact of the graph as a self-consistent Gestalt is too strong and therefore it does not deliver the message it is intended to express.

Confusions take place especially when the original phenomena to be represented are also of a spatial nature.

Let us consider some examples taken from the work of Claude Janvier (1981). The subject is presented with a graph (see fig. 25). He is told that the graph "tells" how the speed of a racing car varies along a 3 km track during its second lap. Some more explanations are added (pp. 114—115).

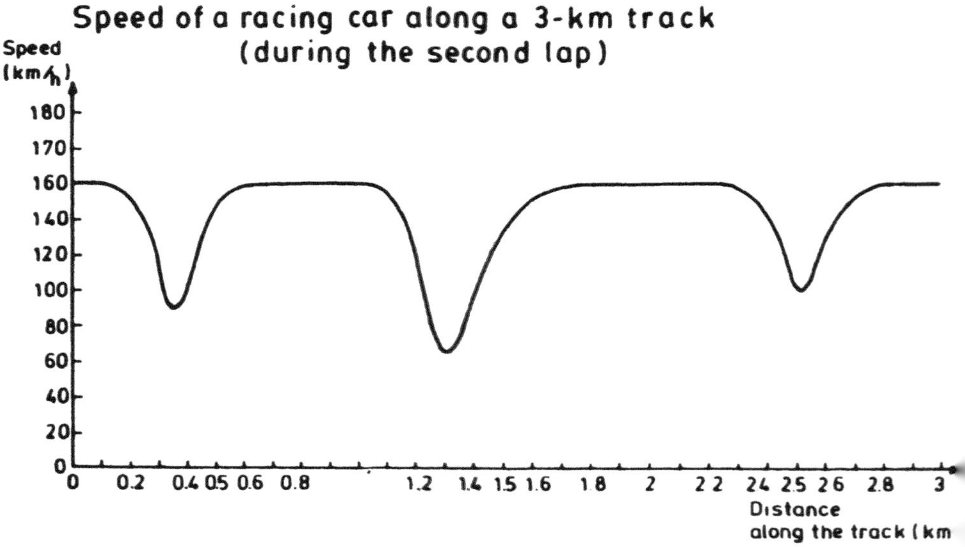

Fig. 25. (After Janvier, 1982.)

The subject is then asked to say how many bends there are along the track on which the car is driven. He is also asked to indicate which bend is the worst, the easiest and the "second worst". If the subject has difficulties he is asked some additional questions each of them providing a hint for the solution. For instance: "What is the top speed? What is the slowest speed? What is the speed at 1 km, at 2.5 km?"

In the second item, the subjects were asked to sketch for each track appearing on the left hand side of figure 26 a speed graph similar to the one presented in figure 25 (p. 116).

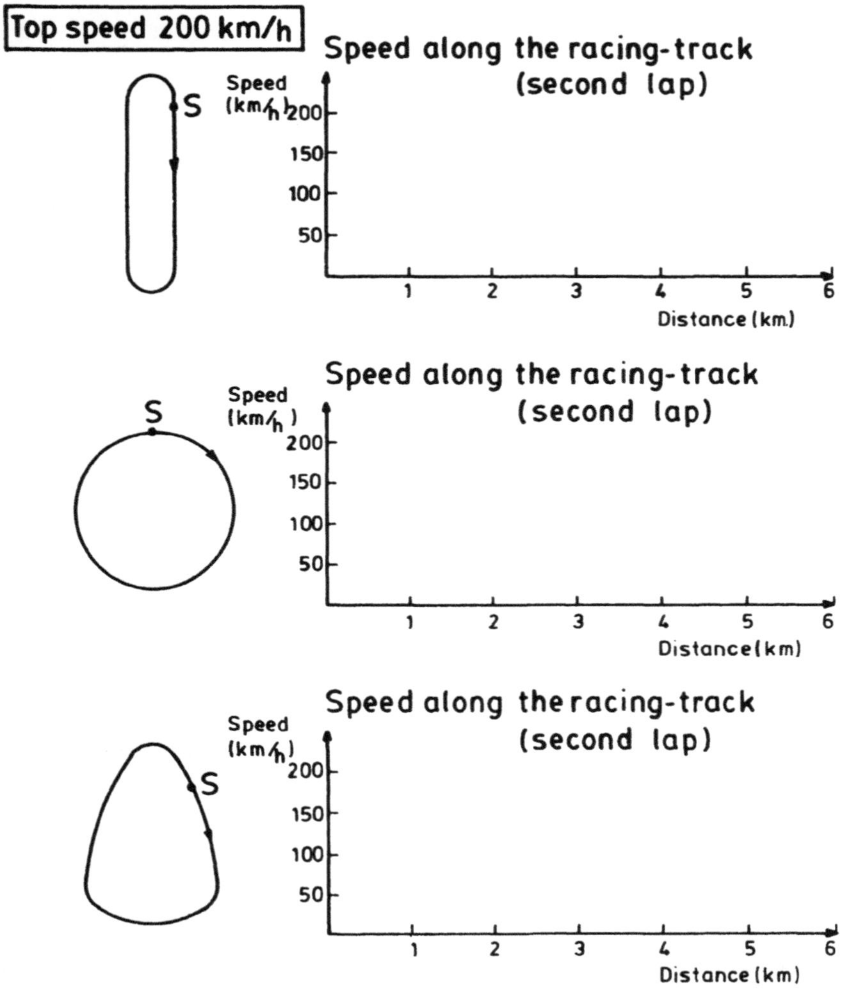

Fig. 26. (After Janvier, 1981.)

A third item asked the subjects to identify among the tracks presented in fig. 27 the track which is represented in fig. 25 (the subject was told that only one track was correct) (p. 117).

The subjects were secondary school pupils aged 11 to 15.

The most common mistake of the subjects, as reported by Janvier, consisted in *confusing the graph with the track*. For instance, with regard to the graph in figure 25, the subjects indicated 6, 8, or 9 bends. "The most fascinating finding", writes Janvier, "was the simultaneous use of the graph at the symbolic and the pictorial levels". It was found that some subjects (12

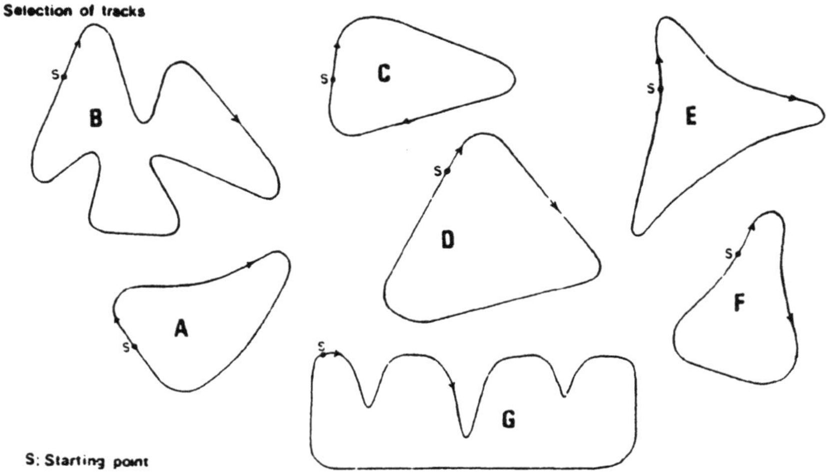

Fig. 27. (After Janvier, 1981.)

girls out of 24 and 2 boys out of 15) were able to correctly interpret the relationships between speeds and points of the graph — the top speed, the slowest speed — but were not able to determine the number of bends in the track according to the graph.

In other words, one may know exactly what each element of the graph symbolizes but at the same time be unable to interpret the graph *as a whole* in terms of the original *as a whole.* This means that the opacity of the graph is not due in this case to a misunderstanding of how the mapping takes place (between the motion of the car and the corresponding graph). It is the structure of the graph, and its figural similarity with the structure of the track which are misleading the subjects.

A second difficulty was revealed by item 3 (the choice of the track corresponding to a graph). In this case the subjects had to consider the relative sharpness of the bends (not only the existence of the bends) in order to be able to distinguish, among tracks *A, C* and *D*, for instance, the one which fits the graph represented in figure 25.

In this case, no intuitive solution is possible at least at the level of experience of the subjects investigated. A step-by-step analysis at a conceptual level is necessary. For instance, in order to decide between tracks *C* and *D* one has to observe that in track *C* the third bend is sharper than the first, while in track *D* the converse is true and this corresponds to the graph. Janvier reports that: "... we observed that the only successful strategy was to give 'verbal tags' to the elements involved and afterwards work basically from those 'spoken' tags." (p. 121.)

Claude Janvier has also remarked that during their analysis, the subjects are frequently misled by pictorial similarities (for instance the association of a straight stretch with a straight portion of the graph).

SUMMARY

What characterizes diagrammatic models is the fact that they represent intuitively the original reality *via an intervening conceptual structure.* Without a clear understanding of this intervening structure, with its laws and constraints, the diagram cannot deliver its message. There is no *natural,* self-evident analogy between the original and the model in the case of diagrams.

Correct analogies may be identified or created but only on the basis of a system of conventions.

If the syntax of the model is not clearly understood by the student, various types of difficulties and errors may emerge. The diagram (the image) may simply impose its own specific, figural constraints — without connection with its modelling function. For instance, the student may not accept that '*A* includes *B*' means either that *B* is contained in *A* or that the sets *A* and *B* contain the same elements. From a strictly figural point of view the two alternatives are not representable simultaneously.

In certain situations, the diagram is so well structured that it may become opaque, simply masking the original. This happened, for instance, in Janvier's experiments where a graph was mistakenly interpreted as being itself the track of the racing car. Very often, only parts of the graph are erroneously identified, by analogy, with segments of the original. Finally, the interpretation of certain graphs may need an explicit, systematic conceptual analysis of their elements and their respective meanings in terms of the original. In other words, a direct global interpretation of the graph in terms of the original (from Gestalt to Gestalt) may be impossible because of the complexity of the data. In this case the role of intuition may be only negative. The step-wise analysis is disrupted and disturbed by the strong natural tendency to proceed according to natural (but incorrect) analogies between the Gestalt of the graph and that of the original.

Diagrams play an important role in scientific investigation. They may play an important didactic role in offering the student visual means for synthesizing data, for representing abstract relationships, for solving various types of problems.

But above all, the student, by learning to produce and utilise diagrams of various kinds, develops his own capacity to take advantage of the enormous heuristic capacities of visual, conceptually controlled, representations.

Janvier quotes Paivio's statement that: "... the retrieval process of encoded information in the memory is controlled by two mediators, one verbal and one imagined" (Janvier, 1981, p. 121).

But a diagram possesses the exceptional quality of offering an intuitive version of the already elaborated, purified conceptual essence of a given reality. By learning to construct and to interpret diagrams the student gets better control over his own mental processes (which is today termed "metacognition") and develops higher-order (conceptually-based) intuitive procedures.

CHAPTER 15

PHENOMENOLOGICAL PRIMITIVES

Andrea diSessa has described a category of models called *phenomenological primitives*, or p-prims for short. These are very closely related to paradigmatic models. A paradigm is an exemplar of a category which is used as a representative for the whole category. It defines for the subject, intuitively, that class of objects. A p-prim is a particular phenomenon which explains or justifies, for the subject, an entire class of phenomena. A p-prim has an *explanatory* function while a paradigm has a *definitional* function. Both act very often in a tacit, covert manner.

The term *phenomenological* indicates that p-prims are not abstract concepts, laws or principles. They are *phenomena* expressed by simple cognitive structures which are monolithic in the sense that they are evoked as a whole (diSessa, 1983a, p. 15). P-prims are accepted intuitively as ultimate, self-consistent explanatory facts. This is the reason why they are called *primitives*. In the words of diSessa:

In the course of learning physics naive students begin with a rich but heterarchical (not being significantly more important than others) collection of recognizable phenomena in terms of which they see the world and sometimes explain it" (diSessa, 1983a, p. 16).

In an unpublished paper diSessa has emphasized some other aspects of p-prims.

P-prims are rather small knowledge structures (involving configurations of only a few parts) that act largely by being recognized in a physical system or in its behavior or hypothesized behavior. In some particularly important cases, p-prims are behavioral, or necessarily entail some associated behavior so that they can serve a self-explanatory role (something happens 'because that's the way things are'). In these cases p-prims become the intuitive equivalent of physical laws, explaining other phenomena, but not themselves being explained, except possibly through their presumably empirical origins. (diSessa, 1983b, p. 4.)

Let me give an example (not mentioned by diSessa). If one asks a naive physics student why an unsupported object falls, the usual answer is: "because it is heavy". Heaviness is a p-prim for the student. It does not need any further explanation. For the naive physics student it is an intrinsic property of objects. It is a "primitive" notion like size or roughness. Heaviness may explain, in the naive student's conception, why objects fall, why they press or pull down the hand which holds them, why a heavy object falls quicker than a lighter one (as he is inclined to believe) etc.

Some of the p-prims may play an useful role in developing the scientific reasoning of the student, since "... they can serve as heuristic cues to specific, more technical analyses" (diSessa, 1983a, p. 16).

The p-prims may, sometimes, become accepted — through a process of abstraction including an expansion of the domain of applicability — as basic, scientific concepts. On the other hand, some p-prims lose their status (as primary notions) because they become explainable in terms of other notions considered scientifically more fundamental.

DiSessa has studied the reactions of a number of undergraduates taking freshman physics courses. Analysing the results of the interviews he was able to identify several interesting primitives with a large range of applications. He tried to imagine the path leading from p-prims of naive students to those of novices and thence to the scientific concepts of experts.

A suggestive example of a p-prim considered by diSessa is that of *springiness.*

The following problem was proposed to a student:

If a ball is dropped, it picks up speed and hence kinetic energy. When the ball hits the floor, however, it stops (before bouncing upward again). At that instant, there is no kinetic energy since there is no motion. Where did the energy go? (diSessa, 1983a, p. 17.)

The correct answer is the following. The ball and the floor compress each other on impact. This compression stores energy like the compression of a spring. As an effect of distortion the stored energy is released and the ball is pushed upwards (*ibid.*, p. 17).

In the above example, the compression and distortion of a spring may serve as an explanatory model for what happened with the bouncing ball.

The behavior of the spring, released after compression, may be accepted intuitively as a natural phenomenon which does not need any further explanation. Intuitively, it makes sense to explain the behavior of the bouncing ball as being similar to that of the spring. Springiness would then be a p-prim for the phenomenon of a bouncing ball. It would be a useful p-prim because, through a process of refinement and abstraction, it may lead to more fundamental notions. Springiness, says diSessa

... is not only consistent with the highest priority (Newtonian) physical ideas, it provides a convenient, organizing conception which frees one from the necessity of always treating spring-like phenomena in terms of idiosyncratic situational details such as how and where exactly physical deformation is taking place. It serves as a macro-model which summarizes the causality (deformation → restoring force → rebound) and energy flow (deforming force drains energy into potential energy which is liberated as the deformation relaxes). (*ibid.*, p. 19.)

A p-prim is, then, a kind of practical theory. Its function is that of lending concrete, organizing support to an intuitive understanding of a system of concepts, which represents the real theory. Like a diagrammatic model, a p-prim model is an intervening device which facilitates, for the subject, the interpretation of a certain category of phenomena in conceptual terms. Usually the expert regards a p-prim as a macro-model which combines the qualities of concrete structurality with that of pointing correctly to general basic notions. Considering, for instance, the bouncing ball as similar to a spring he gets an adequate, intuitive representation which summarizes for him an entire system of explanatory concepts.

A student may sometimes imagine a p-prim which is scientifically inconvenient.

DiSessa describes the attitude of one of his subjects who was asked to explain where the energy is stored when the bouncing ball reaches the floor. She could not think by herself of springiness, and the interviewer suggested the compression of a spring. The subject had a clear intuitive understanding of the behavior of a compressed spring but nonetheless could not accept that the same explanation holds for the ball and generally, for every kind of piece of matter (for instance a ball made of steel). Her justification was that many objects are rigid and then they cannot be squished (deformed). For that subject rigidity and "squishiness" were p-prims, that is to say properties which may be understood intuitively by themselves and which, in turn, may explain other phenomena. But these p-prims are, in diSessa's terms, "of low priority". They are easily explainable through more basic notions (for instance the inter-molecular forces). According to the interviewed subject, rigidity and "squishiness" were two distinct basic primitive properties i.e. two distinct p-prims with high priority.

DiSessa uses the term *priority* with two related meanings. The *cueing priority* of a notion refers to how likely the notion is to be scientifically profitable to the learner. For instance, springiness is much more consistent with a Newtonian world view than rigidity, "so in expert thought it will be used with less provocation than rigidity — it has a higher cueing priority than rigidity" (*ibid.*, p. 20). On the other hand, *reliability priority* is somehow synonymous with being "more fundamental". For instance, concepts like force and energy possess, in the eyes of an expert, a higher reliability priority (possess a higher degree of primordiality) than the notion of springiness. Springiness is explicable in terms of force and energy and not *vice versa*. But reliability priority is context-dependent in the sense that the order of priority of a certain notion is not absolute; it is generally different in a novice and in an expert in relation to the context in which it is used. The student interviewed by diSessa needs to change her understanding of springiness by enlarging the set of contexts which cue the idea. In other words, *the reliability priority attached by the subject to springiness must change in order to explain the bouncing of the ball.* The subject accepts that a spring is springy but does not accept that a ball (for instance a ping-pong ball) may also be springy because, to her, a ping-pong ball is "rigid".

DiSessa imagines a thought experiment in which a novice is learning about potential energy. He imagines his hand pushing the coils of the spring together — "pumping the energy into it". At the same time, he feels that the spring presses his hand as if it wants to expand back. Such an experience may suggest, intuitively, the accumulation of potential energy and its possible transformation into kinetic energy.

Notice how the elements of the interpretation, in particular the set of features to be attended to, are mostly drawn from common sense, and yet the combined effect is to serve as the model of causality and energy storage which is the function proposed in the expert p-prim. The thought sequence binds together in an appropriate way the elements of previous knowledge which serve as basis for the interpretation. The structure of that combination is the new element for the student. The structure is locally justified because it is seen merely as a description of a known phenomenon, the action of a spring. (*ibid.*, p. 29.)

In short, in the given context springiness is a better p-prim of higher priority than rigidity or squishiness because: (a) by itself it justifies intuitively the bouncing of the falling ball while rigidity and "squishiness" do not, and (b) it suggests, intuitively, the fundamental phenomena of storing and releasing energy while rigidity and squishiness do not.

Let us return to our initial example: why does an unsupported object fall? As I said, the p-prim of the naive subject is the "heaviness" of the body. As a matter of fact, scientifically, this p-prim is of low priority. Its explanatory power is very limited. It is, in fact, a dead end in regard to other complex, related physical phenomena. It does not refer to an *external* cause, that is to a *force* which would attract the body and cause it to drop to the ground. It does not justify intuitively the orbital motion of cosmic bodies. The heaviness p-prim, therefore, blocks the way to understanding gravitation which is a fundamental scientific concept.

Let me mention some other p-prims identified by diSessa. Ohm's p-prim involves three components: an impetus, a resistance and an effect. The effect increases with the impetus and decreases with the resistance. It is an intuitively understandable relation with various contexts of application. For instance, "pushing harder to make objects move faster" or the well known law of Ohm, $I = E/R$ (the current flow in a circuit (I) is proportional to the voltage (E) and inversely proportional to the resistance (R). In fact, the above p-prim, although possessing a high priority with a novice, has only a low priority with an expert. It provides only a rapid, qualitative analysis and does not represent the law itself or the scientific explanation of it. Its importance is determined by the generality of its applications and by its intuitiveness (diSessa, 1983a, pp. 21—25).

Force as a mover is an incorrect intuition playing the role of a p-prim in naive and novice students. The idea is that a force causes motion in the direction of the force, ignoring the effect of previous motion (*ibid.*, p. 30; see also diSessa, 1982). For instance, a subject is asked to change the trajectory of a moving body in order to hit a certain target. The subject applies a force to the moving body directly towards the target, "ignoring" the existing motion.

The dying p-prim. This false intuition causes naive and even novice students to believe that a force is always needed to maintain a constant

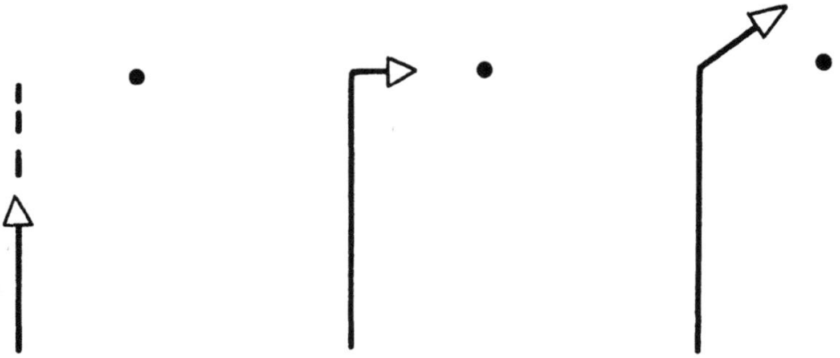

Fig. 28

velocity (diSessa, 1983a, p. 30). It is this intuition which underlies the ignorance of inertia referred to in other chapters.

Another interesting analysis of a p-prim refers to the *musical bell* problem.

Apparently identical bells in size and shape made of the same material may have different pitches. How is it possible? In fact the bells considered (from the Montessori educational material) vary in thickness. The problem is whether the thicker bells are lower in pitch or higher. Most of the subjects affirm that the thicker bells are lower in pitch. This is a very common intuition even in people who do not know that pitch is related to the frequency of vibration. For some more sophisticated students, the explanation is that thicker means heavier which implies slower (lower frequency). The frequently encountered intuitive answer that thicker bells have a lower pitch may be decomposed, by a logical analysis, into the following string of judgments: thicker → heavier → slower → lower frequency → lower pitch.

A few interviewees concluded that *thicker* must mean *stiffer* and, consequently, the thicker bells must have a higher pitch. With regard to expert thought — the underlying model, says diSessa, consists of a perfect spring attached to a rigid support at one end and to a particle at the other. The frequency of oscillation of such a device is proportional to the square root of k/m where k is the spring constant characterizing the stiffness of the spring and m is the mass of the particle. Both mass and stiffness influence the frequency of oscillation, but in an inverse way. But while mass increases linearly with thickness, stiffness increases as the square of thickness. Consequently in the ratio k/m, stiffness wins the competition and then thicker bells have a higher pitch (diSessa, 1983b, p. 15).

In diSessa's opinion, "experts, when they need this information qualitatively, use intuitions of precisely the same form, "heavier implies slower" and "stiffer implies faster", but their confident use is restricted to the specific context of a simple harmonic oscillator. That is, experts attach intuitions of the same kind as those of novices but they restrict them to more specific contexts" (diSessa, 1983b, p. 15). The expert's confidence in the use of such intuitions is "linked to having an elaborate knowledge system which can validate the simple harmonic oscillator" context in any particular case, and can also justify the qualitative results specified by the intuitions with high priority formal notions (a genuine derivation) if necessary (*ibid.*, p. 15).

DiSessa's claim then, is that *intuitions of novices may survive in the expert's reasoning*. They may remain useful in the qualitative forms of this reasoning — although conceptual control intervenes. This claim is of fundamental importance for the theory of intuition, especially bearing in mind that diSessa is an expert physicist.

In the above examples p-prims may be identified which: (a) may be originally correct but which may be used incorrectly in certain circumstances (heavier means slower but thicker does not imply faster i.e. higher pitch); (b) are dead-end intuitions (the role of rigidity versus squishiness in causing a ball to bounce, heaviness as an intrinsic property of objects); (c) are inadequate intuitions (e.g. the dying-away p-prim); and (d) are correct, high priority intuitions which may be acceptable to the expert.

P-prims (elementary physical intuitions) are, in diSessa's view, structural elements of both naive and expert scientific reasoning. They are the constitutive elements of a large vocabulary of configurations through which reasoning is operating (*ibid.*, p. 22). In other words, intuitions participate in a thinking process not as isolated, elementary components but organized in complex structures like sounds in the constitution of words.

A fundamental characteristic of intuitive structures is their resistance to change. At a first glance it may appear difficult to explain the robustness of intuitions. Many of these pieces of knowledge — the p-prims — are easily falsifiable either by logical arguments or empirically. Why then are p-prims so resistant? A first reason suggested by diSessa for this robustness consists in the fact that intuitions are not encoded as explicit propositions but more as "fluid recognitions and expectations" which cannot be easily analysed and possibly rejected (diSessa, 1983b, p. 6).

A second reason is that the replacement system (the conceptual, scientifically valid, knowledge) is organized in vast cognitive structures which are explicitly justified while the p-prims are self-justified, self-evident and directly, locally applicable. Intuitions with their deep roots in the person's experience and their self-explanatory capacity could hardly be replaced by a system in which most of the propositions need extrinsic justification and validation.

Thirdly, although p-prims may be individually fluid they are organized — as has already been emphasized — in coherent structures. "It is through this coherence that priorities are shifted and recorded, and so the system, as a whole, can be much more resistant to change than any individual element" (*ibid.*, p. 6).

DiSessa proposes a very tempting yet hypothetical description of the changes undergone by p-prims as an effect of intellectual development or systematic instruction (from naive to expert physical intuitions).

The development from naive to expert physical intuition is hypothesized to occur in the following ways: First, the rather large, but relatively unstructured collection of p-prims present in naive individuals become tuned toward use in formal physics. Specifically, the priority of some p-prims becomes greatly enhanced or reduced as they find more or less comfortable places in the developing physical knowledge system. Animistic and anthropomorphic p-prims become systematically much less used. Others having closer associations with formal physics, such as some dealing with symmetry and conservation, become generally more used. Undoubtedly some entirely new p-prims are generated as the learner's descriptive apparatus pays attention to different features and configurations in the physical world. But perhaps the most drastic revision in the intuitive knowledge system is in the change in function of p-prims. They can no longer serve a self-explanatory role, but must defer to much more complex knowledge structures, physical laws, etc., for that purpose. Instead, p-prims serve as heuristic cues to typically more formal knowledge structure, or they serve as analyses which do their work only in very particularly defined contexts. One learns when to use force as a mover and when not to (diSessa, 1983b, p. 5).

Do p-prims change themselves or do they only change their role, their priorities, the configurations to which they belong? It seems that diSessa considers the change in role and function but not in the nature of p-prims. Nevertheless, in a different text diSessa writes: "In becoming useful to experts, naive p-prims may need to be modified and abstracted to some extent" (diSessa, 1983a, p. 32).

DiSessa seems, then, to oscillate between these two alternatives, possibly because he does not possess enough experimental evidence for deciding.

In my opinion, there is no such thing as gradual or developmental transformation of intuitions. Intuitions cannot change because they are so deeply rooted in the architecture of our mental schemas that they appear to the subject to be self-consistent, self-evident representations. They are structured so as to offer a maximum of stability, robustness and intrinsic credibility. I do not believe that they are fluid. They are on the contrary "frozen" cognitions. Kruglansky and Ajzen (1983, pp. 31—32) use the term "epistemic freezing" for designating the phenomenon of belief perseverence in both laymen and scientists. I may be absolutely convinced that terms like point, line and surface have only a formal existence, with no objective correspondent. And yet these terms continue to intervene in my reasoning as if they were material realities, though controlled by conceptual constraints.

I know (because I have been taught) that the object in my hand is heavy as an effect of gravitational attraction. Nevertheless, in my every-day mental behavior heaviness continues to be, de facto, a self-explanatory intuition.

Considering Piagetian findings and others referring to developmental phenomena it is reasonable to conclude that intuitions may be replaced by other intuitions *but not transformed.* In the transitional stages, the child may oscillate between two distinct intuitive interpretations of the same phenomenon. But no gradual transformations are taking place. There are no intermediate intuitions between the non-conservation attitude towards quantity, length, cardinality etc. and conservation intuitions. A 7-year-old child considers *a priori* that two equivalent sets of marbles arranged into two rows have the same cardinal even if one of the rows is longer. One may argue that a transformation has taken place from the pre-operational to the operational child but this is an unconscious process. As a matter of fact it is difficult to identify the mechanisms of such hypothetical transformations.

I also consider that the process of production and replacement of intuitions takes place only until the establishment of the formal operational period. If no external, adequate, systematic, long-lasting instruction interferes after the age of 12—13 our intuitive acquisitions remain unchanged. These affirmations are based on our findings referring to the notions of probability (Fischbein, Pampu and Minzat, 1970; Fischbein, Barbat and Minzat, 1971, Fischbein, 1975, pp. 138—135 and 189—201) and infinity (Fischbein, Tirosh and Hess, 1979).

Moreover, we assume that the main progress is in the conceptual control a person is able to exert over his intuitive biases. Our assumption is that our basic intuitions (like those related to space, time, motion, various Piagetian conservations etc.) elaborated during the pre-operational and concrete operational stages can never be totally eradicated (after they have been definitively established). Their coerciveness may weaken, they may become less influential, but they cannot be suppressed. Consequently conflicts may appear between old, strong intuitive beliefs and new, high-priority conceptual or even intuitive structures. I fully agree with the above quoted affirmation of diSessa that: ". . . the most drastic revision of the intuitive knowledge is in the change in function of p-prims . . .". They lose their self-explanatory role which is, then, undertaken by more general basic conceptual systems.

According to Bunge (1962), intuitions may play a positive role in science only at a pre-systematic stage. In order to be useful to scientific reasoning, intuitions have to undergo a process of refinement and abstraction, but in this case intuitions lose their specificity.

> The products of intuition are rough to the point of uselessness: they must be elucidated, developed, complicated. The intuitive "lightning", the hunch, may be interesting if it occurs in the mind of an expert and if it is cleansed and inserted into a theory or at least in a body of grounded beliefs. This is how our intuitions gain in clarity and scope. By being converted into formulated concepts and propositions, they can be analyzed, worked out and logically tied to further conceptual constructions. Fruitful intuitions are those which are incorporated in a body of rational knowledge and thereby *cease* being intuitions. (p. 113.)

This conception is reminiscent to some extent of that of diSessa who has also mentioned a process of abstraction of p-prims. But diSessa insists more on their changing role when incorporated in a higher level scientific reasoning. According to Bunge, the transformation undergone by an intuition in order to become an useful component of a scientific reasoning is so deep that it ceases, in fact, to be an intuition. In contrast, diSessa believes that intuitions do not disappear from a scientist's thinking; instead of having an explanatory function as they have in a naive student, they lose their high priority status and get a heuristic role. One may suppose that Bunge and diSessa differ only with regard to the stage of scientific creativity to which they refer. But my feeling is that the difference is deeper. Bunge strongly believes that intuitions play no role in the construction of abstract theories. "The construction of abstract theories is accompanied by an almost complete elimination of intuitive elements" (Bunge, 1962, p. 114). But the same happens with factual sciences like physics. "Every factual science has tended to achieve higher and higher degrees of epistemological abstraction as it converted given phenomena into problems to be solved. In this sense the progress of factual theories parallels that of mathematics: both become less and less intuitive." (Bunge, *ibid.*, p. 115.)

It is clear that Bunge is projecting features of an elaborated theory onto the process of elaboration itself. When Poincaré, Hilbert, Einstein, Hadamard and, more recently, Paul Cohen, David Tall and Andrea diSessa refer to the role of intuitive factors in scientific reasoning they are certainly considering the psychological dynamics of the process and not the final, purified products. Intuitions and intuitive models are heuristic ones, as diSessa says. They do not exempt scientists from supporting, finally, the validity of their statements by formal or empirical systematic proofs.

The important contribution of diSessa is that he has shown that even in the thinking of highly sophisticated scientists, p-prims (intuitive elementary phenomenological models) do not lose their impact. Scientists also, very often, need concrete, strongly organized apparently self-explanatory models — elementary physical representations of phenomena directly accessible to human intelligence (like springiness, force as a mover etc.).

In contrast to the naive thinker, expert scientists are able to select those intuitive models which are scientifically acceptable (although with a low priority) and to analyze and control them conceptually.

The problem which remains open is that of the possibility of developing *new* p-prims specific to scientists, i.e. new cognitive beliefs as an effect of systematic scientific training. My guess is that such "secondary intuitions", as we call them, are possible. But they do not emerge by transforming primary intuitions — as I have said such a transformation is essentially impossible — but as new beliefs shaped by the constraints of systems of highly elaborated scientific concepts. This is only a hypothesis. It remains to be demonstrated that, for instance, Cantorian, relativistic, or quantum theory notions, or the axioms and theorems of non-Euclidean geometries, may be converted into *beliefs*, like those related in the layman's reasoning to Aristotelian physics or Euclidean geometry.

CHAPTER 16

CONFLICTS AND COMPROMISES

It has been emphasized in the preceding chapters that a major factor in shaping intuitions is represented by what we have termed *intuitive tacit models*. We have mentioned that the heuristic effect of a model depends on its faithfulness to the original, its degree of autonomy with respect to the original and its compatibility with the demands of human reasoning. These constraints may give rise to various contradictory effects. Examples of such contradictions have been encountered in previous chapters; here some further examples will be analyzed.

IMPETUS VERSUS INERTIA

Lucienne Viennot asked 291 university students a question, best illustrated in figure 29. What one sees in this figure are trajectories of balls during their motion after having been launched by a juggler. The balls were at the same altitude at the moment the picture was taken, but their speeds were different. The subjects were asked whether the forces exerted on the balls at that moment were equal or not. More than half of the students answered negatively. The explanation given was that there were two forces: one pushing the ball upwards and another — the force of gravity — pulling the ball down towards the ground. When the ball is launched, it gets a certain amount of force ("capital de force") which is consumed during the motion of the ball upwards. When the "impressed" force becomes equal to the force of

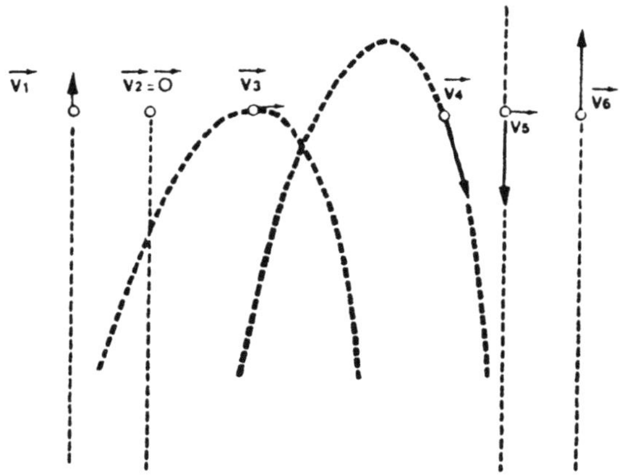

Fig. 29. (After Viennot, 1978.)

gravitation, the direction of the motion changes and the ball begins to fall (Viennot, 1978).

Clement has replicated this type of experiment. He asked engineering students the following question: "A coin is tossed from point *A* straight up into the air and caught at point *B*. On the dot on the left of the drawing draw one or more arrows showing the direction of each force acting on the coin when it is at point *B*. (Draw longer arrows for longer forces)". (see Fig. 30)

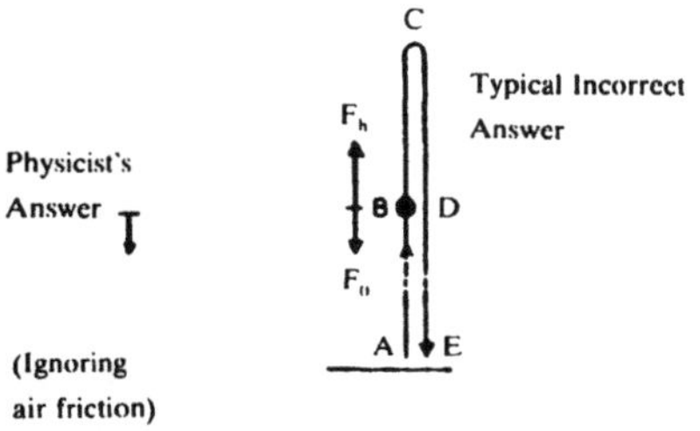

Correct and in-correct answers to coin problem.

Fig. 30. (After Clement, 1982.)

The findings may be considered to be surprising. Incorrect answers were given by 88% of the subjects in a class of engineering students. At the time of the test, they had not yet taken a course in physics. After they had taken the course, 72% gave incorrect answers. The same question was put to engineers who had had two semesters' courses in physics: 70% answered incorrectly. About 90% of the errors involved showed an arrow labelled as a force pointing upwards (Clement, 1982, p. 67). The students, interviewed individually, referred to the arrow as representing "the force to throw", "the upward original force", "the applied force", "the force that I am giving it", or gave an explanation such as "velocity is pulling upwards, so you have a net force in this direction" (Clement, *ibid.*).

As Clement pointed out, Galileo, in *De Motu* (On motion) revealed the same kind of argument: "The body moves upwards, provided the impressed motive force is greater than the resisting weight. But that force, as has been shown, is continuously weakened: it will finally become so diminished that it will no longer overcome the weight of the body and will not impel the body beyond the point" (Galileo, 1960, p. 89).

It was only much later that Galileo changed his point of view and came closer to the principle of inertia. This happened as an effect of his "mental experiments" on motion on an inclined plan (Galileo, 1962). He noted that motion downward on an inclined plane is accelerated whereas motion upward on an inclined plane diminishes its velocity. As a logical consequence, motion on a horizontal plane is perpetual, because there is no reason why it should not keep its constant speed.

As a matter of fact, Galileo came close to the inertia principle but as Allan

Franklin remarked, *he had not reached a complete understanding of it.* To Galileo, "horizontal" meant a surface equidistant from the center of the earth.

The constant speed of a moving body, on which no force acts, means then, in Galileo's conception, motion in a circular trajectory. As is very well known, Newton was the first scientist to give a full, clear account of the principle of inertia. This he stated as the first law of mechanics: A body persists in its state of rest or of uniform motion in a straight line, unless it is compelled to change from that state by forces impressed on it.

But let us consider again, very briefly, the development of the notion of inertia in the history of physics. It is of great interest for the psychology of intuition. It shows how resistant primary intuitions may be, the contradictory beliefs they may produce, the extremely resistant influence an analogy may exert on a conception (especially if that analogy is in fact rooted in the ordinary conditions of our terrestrial life). The principle of inertia was not born as a sudden stroke of genius in Newton's mind, contradicting everything which was accepted before. Certainly, Newton remains the great scientist, who was able to give a brilliant synthesis of the ensemble of data and physical ideas available in his time. But most of those ideas had been envisaged by Newton's predecessors, and some of them had been known since antiquity.

It is a widespread belief that Aristotle's conception was that a body may move only so long as it is pushed by a force. As a matter of fact, Aristotle himself had thought about the principle of inertia as imposed by logical arguments — exactly as Galileo did 2000 years later. "Further", affirms Aristotle, "no one could say why a thing, once set in motion, should stop *here* rather than *there*. So that a thing will either be at rest, or must be moved *ad infinitum* unless something more powerful gets in its way." (*cf.* Franklin, 1978, p. 202.)

This is a clear formulation of the principle of inertia. *But Aristotle could not accept it because it would have implied that his entire system of ideas had to be refuted.* The *intuitive* belief that motion must be sustained by another cause was stronger than the logical, clear conclusion that motion has to continue indefinitely so long as nothing stops it. The logical conclusion was then discarded and Aristotle looked for a different interpretation. When a projectile is launched, what keeps it moving in the medium in which the motion takes place? The medium may act in two ways. It may acquire the power to move the body from the original cause; the power is transmitted from one layer to the next until it gradually dies away. A second moving force is provided by *antiperistasis*; the medium rushes around the body to prevent the formation of a void and this pushes the projectile (Franklin, 1978, p. 202).

Aristotle's theory tended, in fact, to reconcile various relatively independent, and even contradictory, intuitive conceptions. An important difficulty originated in the fundamental belief that every event must have a cause. This determines by itself an apparent paradox. If a body is moving with constant speed one needs a cause to stop it. This would be the primitive formulation of the principle of inertia. On the other hand, if a body is in motion, one must think of a cause which keeps it in motion. This is the view espoused by so-called naive physics. Briefly speaking, we have to admit intuitively that: a body keeps moving if no cause intervenes to change this situation; and a body keeps moving only if a cause (a force) intervenes to preserve the state of motion. As a matter of fact, these two contradictory conceptions are not symmetrical with respect to terrestrial life.

The second conception reflects directly an absolutely consistent practical experience. In the realm of our terrestrial life no motion continues indefinitely without a force supporting it. This everyday experience results in a very firm intuitive belief.

On the other hand, the first conception is a logically derived conclusion based on the general belief that every phenomenon must have a cause. This conception is intuitively weaker and ordinarily masked by the notion that motion must have a cause. The conflict becomes evident only as an effect of an analytical endeavor. The apparently ideal solution of the paradox has been represented in the history of mechanics by impetus theory which simply states that the moving body takes the cause of the motion with it. The impetus theory is somehow a compromise between the two contradictory interpretations. The force is impressed on the body by the original launching act and continues to keep it moving. This force is gradually consumed by the motion of the body. When it is completely consumed, the motion stops. With this theory the paradox disappears. The same cause, "the impressed force", has both effects — it keeps the body moving — as long as there is still a force available, and it makes the body stop when the force dies out.

Moreover, the "impressed force" theory perfectly fits a subjective, practical, behavioral interpretation. If you run or if you pull or push an object you have to use force. You stop when you get tired, that is to say no more force is available. Certainly this theory does not distinguish between notions like force, work and energy. But what is important is the notion *per se* of an active cause impressed on a body which keeps it moving. A main advantage of this — and this represented important advance compared with the Aristotelian antiperistasis theory — was that the "impressed force" theory is compatible with the idea that motion is possible in a void. This rejection of Aristotle's claim — that motion is conditioned by the presence of a medium — opened the way towards the inertia theory. It was also consistent with the atomistic theory stated by some of the great Greek philosophers (like Democritus) which held that atoms could move freely only in an absolutely empty space.

The "impressed force" theory has a long history, changing several times during the period from antiquity to the Middle Ages, with Hipparchus (second century B.C.), John Philoponus (late fifth and early sixth century

A.D.), Avempace (1106—1138) — a Spanish Arab scientist — and St. Thomas Aquinas (1225—1274). All these authors (and others not mentioned) agree on the same basic idea, namely that the motion of a body is possible only as long as a force "impressed" on it keeps the body moving. The discussion which took place over many centuries was not about this point — commonly agreed on — but about the possibility of motion in a void. The general tendency was to reject the initial view of Aristotle which conferred on the medium a propulsive role and to admit the possibility of motion in void.

The entire "impressed force" theory is in fact based on a simple analogical mechanical model. Force is conceived of as a kind of fuel or energy capital (represented, for instance, by heat), which sustains the motion but which is, at the same time, consumed by the motion itself. *Such an interpretation is consistent with the ordinary conditions of our terrestrial life.* It offers a directly, intrinsically acceptable theory and therefore it continues to influence our interpretations (in fact our intuitive interpretations) of motion, even after correct, conceptual knowledge has been acquired.

The most important contribution to the "impressed force" theory during the Middle Ages was probably that of John Buridan (1300—1358) with his impetus theory.

It is not clear whether Buridan conceived the cause of a motion as a force or what has been defined later as the momentum of a motion. In fact, it is not clear either if he saw impetus as the cause or the effect of the motion. In the first case it could have been equated with the notion of force, in the second with that of momentum.

He writes:

> . . . impetus is a thing of permanent nature, distinct from the local motion in which the projectile is moved . . . And it is also probable that just as that quality (the impetus) is impressed in the moving body along with the motion by the motor; so with the motion it is remitted, corrupted or impeded by resistance or a contrary inclination. (Franklin, 1978, p. 204.)

But Buridan also writes: "Hence, a dense and heavy body receives more of that impetus and more intensely, just as iron can receive more calidity (heat) than wood or water of the same quantity". As a consequence, "the iron will be moved farther because there is impressed in it a more intense impetus . . ." (Franklin, 1978, pp. 204—205). The fact that according to Buridan the motion of a heavy body may continue longer than that of a light one (other conditions being equal) seems to suggest that he had in mind some primitive notion of momentum. What is pretty clear is that Buridan, like his predecessors, could not accept that motion may continue indefinitely, without an active factor supporting it. The other ideas he put forward (proportionality of impetus with mass and distance) are, somehow, logical consequences of the basic theory of the self-expanding, impressed force. It expresses the same intuitively acceptable solution to the Aristotelian paradox: a body must continue to move indefinitely if a cause does not stop it (logical implication of the cause-effect intuition); a body cannot continue to move if there is not a motive cause to push the body (a direct implication of practical terrestrial experience). The solution to that paradox, as already stated, is the "impressed force" theory — the body takes the cause of motion with it.

STRIVING FOR COGNITIVE COMPROMISE

The fact that the impressed force theory was a mere speculation, a construct without any experimental basis, and the fact that its authors accepted it as the most natural idea (although it was actually a void notion) *demonstrate again that intuitive consistency is different from factual and logical consistency.* For the sake of intuitive consistency one produces mental surrogates (which may not possess any real, logical counterparts) the role of which is to close down the debate and to eliminate uncertainty, to provide apparent direct credibility and self-consistency for an ensemble of disparate or even contradictory data. *Our point of view is that we are so deeply captured by this kind of conceptual phantasm because it represents a means for protecting ourselves from the paralysing effects (in the behavioral sense) produced by incertitude and incomplete information.*

The process is reminiscent of the "closure principle" of Gestalt theory: with a gap present, there is a state of tension; closure of the gap brings equilibrium in the system. The "gap" may be of a figural, perceptual nature but also of a conceptual one.

Very often, a model is tacitly brought in to help "close the gap", that is to overcome inconsistencies and lack of information. Buridan's model for impetus seems to have been that of heat — an object may store impetus as it stores heat. The advantage of the heat model, compared with the abstract notion of impetus, is that heat has a sensorial meaning and so it confers a sensorial-behavioral dimension on the model.

There are various fascinating examples in the history of science which point to this type of intuitive surrogate intended to reconcile apparent paradoxes. We are not able to imagine the absolute finiteness of reality but we are not able either to imagine the actual, the absolute infinity of any process or of any reality. Aristotle was shocked by that enigma and Kant referred to it as one of the antinomies of pure reason. An intuitively acceptable solution is that of potential or dynamic infinity. This representation does not impose a priori either absolute limits to space and time or the absolute infinity of the universe.

Another solution to this somehow terrifying metaphysical problem is that offered by Einstein. We live in a world which is finite but without limits. A two-dimensional being living on the surface of a sphere would be in such a situation. The area of the sphere is finite but the being we are speaking about could move endlessly without encountering any boundaries. In fact, we live in a four-dimensional, curved space like that conceived by Riemann. Three-dimensional bodies may move endlessly in such a curved space without encountering any boundaries. But the volume of that space would be, nevertheless, finite (Einstein, 1976, pp. 120—125). I admit that the Einsteinian solution does not offer a direct intuitively acceptable representation. But it is less terrifying than Kant's first antinomy and it may offer a model which would not be intuitively contradictory and which may be finally accepted.

Galileo himself, who got so close to the principle of inertia, did not reach a full understanding of it (that is, the acceptance of the intuitively unacceptable representation of a body moving for ever, straight forward, without any motive cause). His solution was also that of a circular compromise: if no cause intervenes to stop it, a body continues to move indefinitely on a horizontal surface — but this, actually, means to move on a curvilinear trajectory. One may claim that Galileo accepted that representation in order to be consistent with the idea that the physical meaning of a horizontal surface, as related to the earth, is that of a curved surface. But I do not think that this is a sufficient psychological explanation. Galileo had enough logical reasons to admit that motion which takes place in the extra-terrestrial space would be rectilinear. But for him, *it was circular motion which was the only perfect one*, the only one which would make possible the uniformity of speed: "The circular motion", says Salviati (and this seems to express Galileo's view)," is first of all finite. The fact that it is finite and uniform explains its unlimited continuity. The rotations are repeating themselves endlessly" (Galileo, 1962, p. 68). Rectilinear motions are only exceptions in nature, affirms Salviati. The natural trajectory of a moving body on which no action is exerted is a circular one.

In my opinion, Galileo's solution to the problem of inertia expresses a compromise, intuitively acceptable to him, between the logical conclusion that a motion, not stopped by an obstacle, will continue endlessly, and the need to avoid the intuitively inconceivable idea of a body moving away and increasing its distance eternally from the departure point. He repeats several times, through Salviati, that rectilinear motions are imperfect, without giving any clear justification.

One may affirm that Galileo succeeded in accepting the idea of motion without impetus and that this was a giant step towards the principle of inertia. But he was still hampered by the troublesome idea of an unlimited space, the necessary setting for such an infinite rectilinear motion, somehow absurdly independent from any possible constraint.

In order to overcome this difficulty, Galileo should have realised, with full clarity, the then unacceptable qualification of uniform, rectilinear motion as a state and not as a process; rest and uniform rectilinear motion are in fact the same physical state. The difference is extrinsic and relative; it depends on the observer's point of view or, in other words, on the system of reference. Such an idea is basically counter-intuitive.

Try to imagine an object free from any influence. Try to understand that, with regard to that object itself, one may equally affirm that it is in a fixed position or that it moves uniformly. One feels that one has reached the limits of intuitive acceptability. As an effect of our terrestrial life we have become used to referring to the ground as an absolute system of reference. We distinguish, behaviorally, between the relative rest of a man sitting in a railway coach and the (pseudo) absolute rest of a man sitting on a platform bench. Intuitively, we accept the idea that a body may be considered either in rest or in motion only in respect to a certain system of reference. But at the same time, it is intuitively very hard to realize the absoluteness of that relativity i.e. *the idea that every point in the universe has the same right to claim that it is in an absolute state of rest while the others are moving* (certainly, the respective point must not be subject to any force).

In short, Galileo was able to reach, on logical grounds, the conclusion that uniform motion continues indefinitely if no force intervenes. This was an essential progress compared to the terrestrial representation of bodies requiring a motion cause in order to continue to move. But there were two other intuitive obstacles which prevented Galileo from attaining the full concept of inertia: the difficulty of conceiving of the eternity of a straightforward motion in an infinite space and the difficulty of identifying absolutely uniform rectilinear motion with a state of rest.

The intuitive compromise was the circular motion: it may continue endlessly, without requiring an infinite space; it synthesizes "process" and "state" in an ingenious manner. The body moves, but, at the same time, it stays in a closed trajectory which dictates the domain of the motion. The fact that real motions of cosmic bodies follow, generally, closed curvilinear trajectories only strengthens the belief that inertially preserved motions must be circular.

The above psychological interpretation is certainly a hypothetical one. But how else can we explain the fact that Galileo, who should have been logically fully motivated (by reasons effectively put forward by himself) to formulate the principle of inertia, failed to reach the final step? It was not a problem of lack of information.

What one may learn from this historical example is the fact that fundamental intuitive obstacles cannot be overcome in the same way as one corrects erroneous information. Such obstacles cannot be overcome by eliminating them one by one. They generally express a whole, complex mental structure which may be actually changed, *only as a whole*, not by pieces. One may consider that Galileo, although he already possessed most of the reasons for accomplishing that kind of mental revolution, was not able to effectively accomplish it because he remained trapped in his intuitive difficulties. It was only Newton who offered the fundamentally new picture, the absolute infinite, homogeneous and isotropic space in which rectilinear uniform motions are identified with states indistinguishable from the state of rest and in which forces are causes of changes of the state of a body (see Piaget and Garcia, 1983, pp. 213—214).

The history of the concept of inertia is reminiscent of that of negative numbers. As has been seen, great mathematicians like MacLaurin, Euler, Laplace, and Cauchy, although possessing all the conceptual ingredients for defining precisely negative numbers and operations with them, kept struggling in a welter of intuitive difficulties. It was Hankel who accomplished the revolution. He simply rejected any reference to intuitive interpretations. He did not look any more for an intuitive meaning to make negative numbers acceptable. Rather, he formally defined the numbers and the operations with them so as to represent a logically consistent conceptual system. But this was, like the concept of inertia, a totally new philosophy. Both examples point to the same type of situation. In the history of scientific thought new facts and conceptual constructs are accumulated in various domains, creating the need for reforms of conceptual structures and general interpretations. *But before things become clear and settled there is a long transitional period, which may cover centuries, in which the struggle is mainly of an epistemological nature; fundamental intuitive obstacles simply delay the scientific community reaching the essentially new synthesis although all the basic conceptual ingredients are already present.*

The accomplishment of the revolution does not mean that the intuitive obstacles have been definitively eliminated. They probably continue to survive even in the expert's mind though only as subjective, personal, conceptually controlled difficulties. Such obstacles certainly continue to play an active role in the interpretations of the layman and very often in persons who have already acquired the corresponding scientific knowledge.

IMPACT OF THE IMPETUS MODEL: FURTHER EXPERIMENTAL EVIDENCE

Recent psychological investigations have stressed again the impact of the impetus model on thinking about physics. (See, for instance, McCloskey, 1983; McCloskey and Kohl, 1983; McCloskey, Washburn and Felch, 1983.)

In the McCloskey, Washburn and Felch experiment the subjects were presented with the picture of a walking man holding a metal ball. They had to indicate the path of the falling ball (fig. 31). When the ball is released by the walker (assuming that he continues to walk) the subjects had to choose among three pictures: the first indicating that the ball will move forward describing a parabolic trajectory; the second, a straight-down path; and the

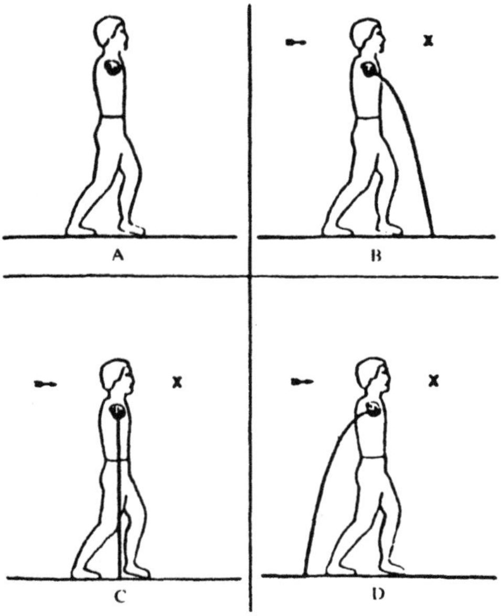

The walker problem (A), the correct response (B), the straight-down response (C), and the backward response (D).

Fig. 31. (After McCloskey *et al.*, 1983.)

third, a backward path. The subjects were 99 undergraduate students. Thirty seven of them had never taken a physics course whereas the others had completed at least one high-school or college course. Only 45% of the subjects indicated the correct forward parabolic trajectory; 49% affirmed that the ball would fall straight down; 6% chose the backward solution. Even among the students trained in physics 40% chose the straight down solution.

In the above example the object, the ball, was carried by the walker. It did not move "on its own" before falling. What would the subjects answer if a body was moving "on its own" before falling?

In fig. 32A there is a schematic representation of a canyon with a conveyor belt (the horizontal line constructed above it). To the belt is attached a metal ball held by an electromagnet. At a certain moment the ball is released. The subjects' task was to draw the path of the falling ball after it had been released (air resistance was to be ignored).

In figure 32B the diagram represents a side view of a canyon with a level ramp extending over the edge. A metal ball placed on the ramp is given a push. The ball rolls along the ramp until the end of it and there the ball falls down. Here, too, the path of the falling ball is requested. In both problems the texts indicated the same distance from the edge of the canyon to the

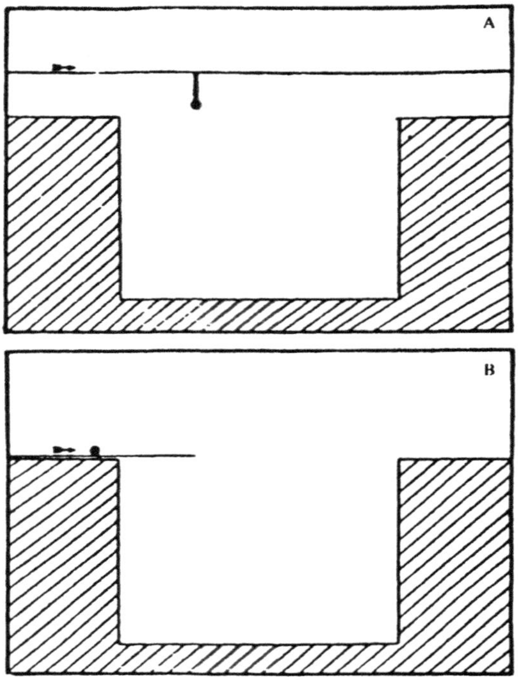

The conveyor problem (A) and the ramp problem (B).

Fig. 32. (After McCloskey *et al.*, 1983.)

point where the ball is dropped and the same speed when the ball begins to fall. For the conveyor problem 65% of the subjects gave forward responses, 23% gave straight-down responses and 13% chose the backward path. In the ramp problem (the ball was supposed to be moving "on its own") 94% indicated that the ball would move forward as it fell and only 6% indicated the straight-down path.

Objectively, physically, in both situations the trajectories should be identical. But from the impetus theory point of view there is a fundamental difference. In the first problem the ball is carried by the belt. It does not move on its own. It has no "impetus". Consequently, on starting to fall it should move straight down. In the ramp problem, the ball prior to falling moves "on its own". The initial push has charged the ball with a certain "impetus" which will confer a horizontal component on the trajectory of the motion.

Considering only the answers to the ramp problem we might suppose that almost all the subjects were using correctly the *inertia* principle. But the 23%

of straight down responses to the 'belt' problem prove that, in fact, the impetus interpretation is the real model for many of them.

As we mentioned above, impetus expresses a compromise between two contradictory constraints: the idea — logically justifiable — that a body keeps moving if no cause intervenes (there is no change without a cause); and the empirically based intuition that a body stops moving if the motion is not sustained by a cause.

But, as we have previously emphasized, the compromise is imperfect. The effect of impetus, unlike that of inertia, is limited in time. The impetus interpretation is imperfect also from an empirical point of view. The subjects should have learned from experience that carried bodies continue to move in the direction of the motion after being released.

What happens is that *in real empirical conditions the subjects are not completely faithful to their theoretical predictions.*

In one of their experiments McCloskey, Washburn and Felch have asked the subjects to actually release a ball, while walking, so as to hit a certain target. After the experiment was performed the subjects were asked about their intentions. Did they intend to drop the ball before, directly above, or after the target? Out of 21 subjects (undergraduate students) seven declared, when interrogated, that they intended to release the ball directly over the target in order to hit it. Actually, six of them released the ball *before* the target (correctly), not right over it as their declared intention was. This means that the practical behavior of the subjects conflicted with their impetus theory somehow compromising with the objective constraints of the given situation.

CONCILIATORY MODELS

In both examples mentioned above, that of the concept of inertia and that of negative numbers, we may identify similar difficulties. In both cases a certain concept, although imposed by logical, formal reasons appears to be counter-intuitive. It does not correspond to the basic intuitive representations people have and this is because the respective logical constraints (negative number, inertia) seem to be behaviorally, practically unsuitable. A number expresses behaviorally a multitude or a measure. An object needs a force in order to move. Nothing is, in reality, less than nothing and a body does not move unless a force is acting on it — according to our fundamental, natural understanding of reality.

In order to overcome the conflict, one looks — implicitly or explicitly — for a *model* which would be able in a somehow sensible manner, to reconcile the conflicting constraints. The model has to be internally consistent, intuitively (directly) acceptable and manipulable. But it has also to reconcile, in one representation, the sometimes contradictory demands of logic and practical experience.

The impetus theory has been an incomplete solution for the inertia problem. It explains how a body may keep moving without a force acting on it — as an effect of an "impressed" force. It is incomplete because, in fact, the logical construct of an *eternal* motion — if no cause intervenes — is not implied by the impetus theory. The impetus is something which is "consumed" by the motion like a fuel. A better, more ingenious compromise was that of Galileo. The imaginary impetus was eliminated. Circular motions are indeed eternal, but "kept under control". They are eternal if no cause interferes but, nevertheless, they may be entirely assimilated to (static) states.

Let us briefly recall what was said about negative numbers.

Negative numbers have been known from antiquity. They appeared as formal products of mathematical operations. They were intuitively and practically unacceptable. In the history of mathematics, various interpretations, various conciliatory models have been tried. It has been affirmed that a negative number is a "transitory" entity. But the term "transitory" has a realistic meaning, not a mathematical one. In mathematics, an entity exists or does not exist; in the former case, an existence proof is needed. Others have claimed that a negative number is an inverse version of a genuine number. Consequently a negative number can never exist by itself but only as a kind of reflection of a "true" number. Such a claim also has no mathematical meaning. A concept gets a mathematical status if it is formally defined and the operations with it are also formally defined, without leading to contradictions, in the realm of a certain axiomatic system. The image of an object in a mirror exists only as long as the object and the mirror exist. In mathematics such realistic, physical relationships have no meaning. The impetus model, the "circular" interpretation of inertia, the "transitory" and the "reflective" interpretations of negative numbers were, more or less, internally consistent, relatively autonomous representations of the respective concepts and phenomena. They offered intuitive, mentally manipulable, intrinsically meaningful representations. Certainly, they could not account consistently for all the facts they were supposed to represent. They were compromises. They were imaginary creations, the role of which was to offer, in such contradictory circumstances, internally consistent, intuitively manageable constructs. Our hypothesis is that, *when facing contradictory data, the mind tends to produce, automatically, such conciliatory*, stabilizing devices. Their adaptative function is to secure the continuity and the current productivity required by an efficient mental activity although they may be unacceptable from a strictly scientific point of view.

Various examples of such tacit compromising models may be quoted in both scientific reasoning and in layman interpretations. "Dynamic" infinity is an intuitively acceptable representation for the intuitively unacceptable notion of infinity. The statistically governed "cloud" of particles is more easily acceptable intuitively than the contradictory representation of particles as possessors, at the same time, of both corpuscular and wave characteristics.

The timeless definition of limit through the inequality $|a - a_n| < \varepsilon$ is mediated by the intervention of a dynamic metaphor.

Recall the definition: "The sequence $a_1, a_2, a_3, \ldots$ has the limit a as n tends to infinity, if, corresponding to any positive number no matter how small, there may be found an integer N (depending on ε) such that $|a - a_n| < \varepsilon$. (Courant and Robbins, 1969, pp. 192—193). Strictly speaking, the definition, being a mathematical definition, should be conceived as timeless. The term "tends" should not be conceived as a real dynamic process of producing successive numbers in which the end of the process would be the attainment of the "limit". Such a productive activity may take place in the realm of an empirical reality (for instance cells which multiply themselves) but not in mathematics. The numbers a_1, a_2, a_3 exist. They are given. The definition of limit describes the relationship between this *given* infinite set of numbers and a certain number termed "the limit". Mathematics has to do, strictly speaking, only with relationships, never with input—output processes. But the dynamic, time-governed, metaphor is so deeply rooted in our reasoning activity that we simply cannot get rid of it. *The term "tends" expresses a compromise between the dynamic, intuitively meaningful representation of a travelling process (which takes us from A to B) and the abstract, Platonic representation of a given relationship.* In reality, objects do not "tend": They rest or they move. *We* tend as living beings. The "to tend" metaphor expresses a subjective feeling not an objective event. A bullet moves in the direction of a target. It does not "tend" to the target. "Tending" implies having an anticipatory representation of the target.

The concept of limit was originally related to a motion metaphor obviously because calculus was originally, the study of change. According to Kaput (1979, p. 295): ". . . for very good reasons, generations of brilliant mathematicians struggled to attain logical control over clarification of the meaning of this metaphor, leaving us with the (timeless) conditional statement as the ultimate definition". In fact, the success was not complete. The expression "the numbers tend to", and the arrow representation, are reminiscent of the motion interpretation. The difference is that what "tends" does not necessarily move. It stays but it moves potentially. *The "tends" metaphor is the intuitive compromise between the subjective constraint to consider the "limit" concept in a dynamic context and the mathematical constraint of not accepting time and motion in the frame of its own axiomatic system.* For defining formally the limit concept we cannot follow the natural course of thinking. Naturally, we should have in mind first the independent variable and then the dependent variable. "But this natural attitude is not capable of clear mathematical formulation. To arrive at a precise definition we must reverse the order of steps" (Courant and Robbins, 1941, 1969, p. 292). That is to say, we consider *first* not the approaching process but what *should* happen — as an effect of it — to the dependent variable.

Reversibility is a fundamental quality of logical reasoning as Piaget has shown. Logically, formally, equivalence is an absolute symmetrical relationship. The logical inversion required by the definition with ε of the concept of limit should not, in principle, raise any problem. The fact that one faces initially great difficulty in grasping it is more support for the argument that logical, formally accepted principles do not solve the problem of genuine understanding. "Our intuition", affirm Courant and Robbins, "suggests a 'dynamic' idea of limit as the result of a process of motion. We move on through the row of integers 1, 2, 3, . . . , n and then observe the behavior of the sequence A_n. We feel that the approach $a_n \rightarrow a$ should be observable." (Courant and Robbins, *ibid.*, p. 292.)

As we have already said, the compromise metaphor represented by the "to tend" notion survives even in the structure of the apparently pure, formal definition (with ε) of the concept of limit. Even if we put the cart before the horse (the dependent before the independent variable, as logically imposed in this case) we are still bound by the psychological constraint of keeping alive the idea of a process when considering the mathematical notion of limit.

It is, then, no wonder that the student is totally perplexed when first learning (without appropriate preparation) the formal definition of limit.

In our opinion, as we have frequently stressed, by simply ignoring the intuitive constraints in teaching mathematics (even advanced mathematics), one does not solve the didactical problems. The conflicts between the logical and the intuitive constraints tend to survive as latent contradictions and manifest themselves in non-standard situations; or they may generate compromises carried by apparently harmless metaphors embedded in formal propositions.

It often happens that the expert and the layman or the teacher and the pupil may interpret the compromising representation in different — even opposed — ways: For the teacher, the metaphor "a_n tends to a" does not mean that n effectively takes every value (in natural numbers) from 1 to infinity but rather that the difference $|a - a_n|$ may be as small as one wants. In contrast, the pupil tends (!) to see in the metaphor "tends to" an effective process of numbers running in a certain direction.

Let me add another example. Many students when asked about the value of 0.333 . . . do not accept that that symbol is equivalent to 1/3 (0.333 . . . = 1/3) but only that it tends to 1/3, that is to say that it necessarily remains smaller than 1/3. The student sees in the metaphorical image of the set of three points the dynamic process (practically never finished) of approaching 1/3, while the teacher accepts the (non-intuitive) convention that the symbol . . . means the *actual*, denumerable infinity.

It is no wonder that the student is perplexed by all these metaphors which, in fact, hide inherent contradictions rather than resolving them.

The basic didactical approach which, in our opinion, may help the student to cope with such situations is to make him aware of his own intuitive constraints and of the sources of the mental contradictions presented by the teacher, as general, human difficulties when handling the respective abstract notions — as they really are.

SUMMARY

One of the main mechanisms in shaping intuitions is that of producing compromise models for intuitively conflicting representations. For Galileo, a circular, "perfect" motion represented a compromise between the need to admit the infinity of a motion if no cause stops it, and the intuitive need to avoid the unacceptable representation of the infinity of space and time. The impetus theory is another compromise for a related difficulty; an "invested" force explains why a body may go on moving without apparently being pushed and, at the same time, explains why eventually it slows down and stops — the "invested" force has been exhausted. The concept of dynamic infinity represents a compromise between the finiteness of our mental schemas and the actual infinity of mathematicians. It is expressed in the definition of limit by the word "tends"; the values, the numbers "tend", like living entities, to a limit. Such conciliatory models are, generally, produced automatically so as to best fit the demands of our mental activity. Usually, they are "representable", they have a behavioral meaning, they correspond to the practical requirements of our terrestrial life. Such compromises take place not only between two intuitive contradictory tendencies but also between logical and intuitive constraints or between intuitive and behavioral constraints.

Formally, the concept of negative numbers is inspired by the negative results of certain mathematical equations. Practically, intuitively a negative number has no meaning. The intuitively acceptable compromise is that of symmetrical opposed magnitudes. The number -5 represents, in an intuitive compromise, a magnitude having an opposite practical meaning in relation to another magnitude defined as positive (negative and positive bank balances).

From the existing evidence one may conclude that such conciliatory models are very robust. It is natural that they should be so. They are intended to preserve the mental equilibrium of the individual. They have "closed down" the debate, the conflict, in a practically satisfactory manner. To eliminate them would mean returning to a state of mental conflict which would have a disturbing effect on mental activity.

But such models, although satisfactory within certain limits, tend to block the attainment and the acceptance of higher order, more comprehensive and adequate solutions. Buridan's impetus model and the circular "perfect" motion model of Galileo blocked the full understanding of the concept of inertia. The "mirror" model of symmetrical, opposite magnitudes blocked the correct understanding of the concepts of negative and directed numbers.

In order to overcome such intuitive obstacles one has to become aware of the conflict as such and to create fundamentally new representations. The conflict which existed at the intuitive level and which the compromise model has been intended to overcome disappears at a higher order conceptual level. In physics a new system of principles and laws, in mathematics a new axiomatic structure, should eliminate the contradictions. The principle of inertia is not a compromise between two contradictory representations. It states clearly the identity between the state of rest and the state of uniform rectilinear motion. This is a counter-intuitive representation but it is an organic component of a logically structured conceptual system — Newtonian physics. Certainly, a fundamental conflict remains between the Newtonian conception and the compromise model of naive physics.

I do not consider that naive intuitive interpretations may be eradicated altogether. The problem is to help the student to understand the logical structure of the new conceptions and their superiority expressed in higher internal consistency and higher order comprehensiveness. This is a new type of conflict generated by intellectual education. It is no more a latent conflict between partial incomplete representations but rather an open conflict between rudimentary intuitive representations and a higher order conceptual structure. It is possible that, in the long run, the new logically-based interpretation will generate a new intuitive acceptance — *a secondary intuition.*

CHAPTER 17

FACTORS OF PERSEVERANCE AND CLOSURE; THE PRIMACY EFFECT

One of the main characteristics of intuitions is their resistance to change, their reluctance to admit alternatives. Kruglansky and Ajzen (1983) refer to this phenomenon as "epistemic freezing", stating that:

> ... belief perseverance reflects the phenomenon of epistemic freezing whereby the person ceases, at some point, to generate hypotheses and accepts a given currently-plausible proposition as valid. Epistemic freezing is considered to be an inevitable feature of the judgmental process because of the potentially endless character of cognition generation. The epistemic sequence must come to a halt at some point lest the individual be left without any crystalized knowledge necessary for decision making and action. (p. 31.)

In principle, the same idea has been expressed in the first chapter of this book. It is only a difference in terminology; we call the frozen beliefs intuitive cognitions.

What are the mechanisms by which the perseverance of intuitions is achieved?

A basic factor already mentioned by us (mainly relevant for ground intuitions) is that of experience. Basic intuitions are shaped by a long-lasting experience, by steady, practical, behavioral conditions. Experience, then, implies a certain group of constraints and requirements, but at the same time it offers numerous opportunities for confirming and reinforcing the corresponding beliefs. As has already been said, such beliefs become an integral part of a large segment of our current behavior. Renouncing them would upset the whole structure and the equilibrium of our behavior. We cannot cease to belief intuitively, behaviorally in the properties of "flat" Euclidean space, for instance, because all our spatial reflexes are adapted to that framework. These ideas have already been discussed and I shall not dwell upon them here; rather I will mention some other aspects.

A second factor related to the previous one refers to the nature of our mental schemas. Each period of intellectual development is characterized by the emergence of a certain number of fundamental intellectual schemas. The various types of conservation, for instance, are deeply related to the specific operational/equilibration capacity of children starting at the age of about 6—7. Internal regulations, the closure property of a certain system of operations, reversibility, and decentration are all conditions of the development of various categories of conceptual invariants expressed in the act of conservation. The conservation of quantity and that of cardinality are associated with the feeling that one deals with properties which are a priori invariant in the given circumstances. But this "a prior" is a consequence of the functioning of the ensemble of mental schemas reached by the child at a certain age.

The stability of intuitions is related to the stability and internal coherence of the mental schemas. As Piaget has frequently emphasized, mental operations are always organized in large systems working according to certain schemas. A number belongs to an infinite collection of numbers. The concept of number, in turn, is based on schemas like classification and order.

Intuitions are organized according to the same mental schemas as operations. Their specificity consists in their implicitness and in their tendency to attain the appearance of closed and self-consistent cognitions. These properties are obtained chiefly through a kind of automatization and condensation of the underlying mechanisms.

The resistance of intuitions to alteration expresses the tendency of the respective systems of schemas to conserve their unity, their hierarchy and their behavioral role. As already said, it is through a lasting experience that such systems of schemas are built and repeatedly reinforced.

The primacy effect is a factor which may also contribute to the perseverance of certain intuitions. "Primary effects in judgmental behavior are generally said to exist when, in judging an object or a person, the individual bases his/her inferences predominantly on early information and appears to be affected less by late information" (Kruglansky and Freund, 1983, p. 452; see also Luchins, 1957).

For instance, for various reasons a teacher may form an initial negative impression of one of his pupils, a physician may establish a preliminary diagnosis on the basis of a group of symptoms, a scientist may produce an initial interpretation of a group of findings. The primacy effect means that the fact itself of being the first — the first impression, the first interpretation etc. — influences the perseverance of the subject's initial attitude. The individual tends to favor, to push forward this first global interpretation, sometimes against other alternatives and even in disregard of sensible arguments or later information which contradicts the first decision.

In the words of Kruglansky and Freund: ". . . an individual may attain closure early in the informational sequence and be relatively impervious to later information" (Kruglansky and Freund, 1987, p. 452).

Let me mention an experiment carried out by Kruglansky and Freund.

Eighty school students were asked to predict the success of a certain person as the president of a company. In the high-evaluation-apprehension condition subjects were informed that they had to predict the success at work of the candidate. In the low-evaluation-apprehension condition, the subjects had to make a similar prediction but they were informed that the selection method used was only at a pilot stage.

One group of subjects were given initial positive data about the target person (for instance his attitude towards the employees' performance and towards their welfare) and subsequently negative data about the same candidate. Another group received the same information in a reversed order (negative-positive sequence). A third variable included was time pressure. In the high-time-pressure condition, subjects were informed that they were allowed 3 minutes to complete their predictions, while in the low-time-pressure condition, they were informed that they would have an unlimited time for solving the task.

The main finding of this research was that judgments in the positive-negative sequence were generally more positive than those in the negative-positive sequence. This suggests that *a primacy effect* was operating. Primacy effects were significantly more pronounced when time pressure was high and when the evaluation apprehension was low (Kruglansky and Freund, 1983, pp. 452—454).

It seems reasonable to infer that the priority of a certain impression is a factor which influences its stability (even against subsequent contrary evidence).

It is worth emphasizing that the primacy factor is effective mainly under time pressure or when the solution requested is presented as not having a decisive importance. This might suggest that the primacy effect is stronger in circumstances in which the individual's critical, selective, attitude is weaker.

In short, the Kruglansky-Freund experiment shows the existence of the tendency of individuals to generate a global impression on the basis of an initial amount of information, to fix that impression as an apparently coherent structure and to render it relatively impermeable to later evidence.

Premature closure is certainly one of the basic mechanisms of intuitions.

Two aspects are to be considered in this context. We refer first to the concept of closure itself — an important concept of Gestalt psychology. One tends, naturally, at the perceptive level to "close" a perceived figure, to complete gaps if there are any, in order to get an image which fits already familiar schemas, which is apparently self-consistent, which is manipulable as it is, as an object, as a whole. It is hard to condense in a global, synthetic meaning conglomerates of discontinuous information. Therefore we tend automatically to fill in, intuitively, the gaps, to "close" the figure and reduce the uncertainties.

The same happens with ideas. We tend automatically, naturally to dramatize fragments of information (in the sense in which Freud considers dramatization of pieces of reminiscences of a dream) and to confer on them a unitary meaning. The closure effect is related to the manipulability of thoughts, to the self-consistency of cognitions and to the natural tendency towards internal equilibration of mental operations. Open problems are troublesome, intriguing.

With this we come to the second aspect relating to the primacy effect; premature closure.

One tends not only to close the search for arguments, one tends to close the debate *as early as possible*. In other words, the phenomenon of closure does not explain by itself the primacy effect. The subject could wait until he had gathered all the information needed and then close the debate. But it is not so. Very often the moment of closure takes place prematurely on the basis of incomplete information. *This means that the need for being able to take a firm decision is stronger than the need to know.* The primacy effect may then be explained by the same fundamental need to reduce as much, and as early as possible, oscillations and hesitations in our current behavior in order to provide the decision mechanisms with (apparently) firm grounds despite the incompleteness of information.

What can explain the perseverance, "the freezing" of these initial intuitions? Kruglansky and Ajzen refer to the following aspects:

The need for structure which expresses "a desire to have knowledge on a given topic, any knowledge as opposed to a state of ambiguity". In other words it is the need for structured information itself which exerts an inhibiting influence on the hypothesis generation process "because the generation of alternative hypotheses endangers the existing structure" (Kruglansky and Ajzen, 1983, p. 16).

The fear of invalidity would, on the contrary, contribute, in the view of Kruglansky and Ajzen, to looking for alternative hypotheses. The subject fears to commit himself to a potentially erroneous hypothesis (and therefore he may manifest the tendency to consider additional possibilities).

In fact, in our opinion, the fear of invalidity is very often weaker than the fear of invalidation. In order to preserve the already reached equilibrium (as an effect of closure) the individual tends to avoid the situation of facing different alternatives.

A third aspect mentioned by Kruglansky and Ajzen is *the preference for desirable conclusions*: "... individuals are more likely to reach conclusions congruent with their wishes than incongruent ones" (*ibid.*, p. 16).

One may assume that the premature closure which characterizes the primacy effect is influenced by the individual's system of values, tastes, interests, preconceptions etc. Accordingly, *the factors which have determined the first judgment will determine also whether the individual will persevere in his initial judgment.*

In fact, the problem of the perseverance of intuitions is much more complex. In order to accept a cognition as self-evident, as indisputable, the individual has to identify himself intellectually, behaviorally and even emotionally with the respective conceptions. Despite the fact that I know formally that the universe is infinite, without privileged directions, I still behave as if I believe intimately, tacitly that the ceiling representation of the sky is an objective reality situated far away over my head. A religious person — even if he is a scientist — when praying to God, looks upwards or at least considers the object of his prayer to be situated somewhere above his head. One can give up more or less easily a non-evident, formally proven idea because its justification does not belong to our natural, essential way of thinking. But a self-evident idea is *ours*, is identified with our basic beliefs, and then it cannot be abandoned. The fact of admitting that we were wrong when we accepted as self-evident a certain notion would be equivalent to doubting the basic mechanisms of our reasoning production and control. Explicitly, formally we may claim that conceptual, scientific knowledge is superior, but tacitly the process of abandoning an apparently self-evident conception is deeply troublesome if not impossible.

On the other hand we tend to be committed to any conception we have enunciated. By defending our decision we defend the quality of our expertise itself — it becomes a problem of personal prestige. In order to defend a conception successfully we have to believe in it. Thus the individual becomes more and more involved in his primary hypothesis. To its conceptual nature a new dimension is added; it becomes *a belief*, an apparently self-evident truth.

Let us now consider some developmental implications of the primacy effect.

We have already mentioned the question if the perseverance of a primary intuitive option is explainable by the mere fact that it was the first intuitive decision. Another possibility is that the choice was made for certain identifiable reasons, and that these reasons explain both the primacy of the given attitude and its tendency to persevere.

Let me recall an example. We have formulated the hypothesis that many of the typical mistakes made by pupils in solving elementary arithmetical problems can be explained by the existence of certain primitive, implicit intuitive models associated with the arithmetical operations.

For instance, we have assumed that the primitive, implicit model for multiplication is repeated addition. If the numerical data of a problem contradict the constraints of the corresponding model, the subject faces difficulties in finding the right solution.

The "repeated addition" model is certainly the first interpretation of multiplication the child has encountered. Is this model so perseverant, simply because it was the first encountered or the first offered by the teacher or has it certain intuitive and didactical properties which also account for its perseverance?

The question is complex and we do not possess enough experimental evidence for a clear-cut answer. Our suggestion is that, in this case, the primacy of the model is determined not by a chance encounter but by deeper reasons. Teachers generally use the repeated addition model for teaching multiplication in the elementary classes because this model fits certain specific properties of the children's reasoning. (We are referring here to the concrete operational period.) Addition is a first-level operation, and children are able to grasp its meaning directly and behaviorally. Multiplication, in its general form (including, for instance, multiplication of fractions and decimals) is a second-degree operation governed by formal rules generally with no intuitive meaning.

One may then, plausibly, affirm that teachers are using the repeated addition interpretation for initially teaching multiplication because this is the natural way for children to understand multiplication. Moreover, even in adults it is the natural way to understand multiplication intuitively.

Consequently, the first interpretation of multiplication encountered by the child is not the first by mere chance. It is determined by the specificity of the child's mental schemas at the respective developmental period.

We may then assume that the primacy effect — the perseverance of a certain initial representation — may be explained not only by its being the first but also by its fitness to the basic requirements of intuitive cognitions.

In others words it may be assumed that, at least in certain conditions, initial conceptions survive for a very long time not merely because they represent a first experience but because they were chosen from the beginning so that their stability would be guaranteed (in accordance with the individual's intellectual features).

A scientist who has formulated a certain hypothesis did not formulate it by chance; it optimally suits his general philosophy in the given domain, his usual way of interpretation, his knowledge, and his research methodology. He is certainly very anxious to preserve his initial interpretation not only for his own prestige — which is certainly an important factor — but chiefly because it is the hypothesis which is best integrated in the structure of his reasoning. He will be unwilling to give up this first hypothesis because by renouncing it he has to re-evaluate a whole system of conceptions.

SUMMARY AND COMMENTS

The primacy effect determines the survival of global, initial impressions, interpretations or decisions, sometimes even against subsequent contrary evidence. It is possible that the primacy effect is influential by itself in generating intuitive beliefs. But it is also possible that other factors may intervene such as specific intellectual constraints and personal commitments.

From the educational point of view there is an important problem to be considered by curricula writers and by teachers. A certain interpretation of a concept or an operation may be initially very useful in the teaching process as a result of its intuitive qualities (concreteness, behavioral meaning etc.). But as a result of the primacy effect that first model may become so rigidly attached to the respective concept that it may become impossible to get rid of it later on. The initial model may become an obstacle which can hinder the passage to a higher-order interpretation — more general and more abstract — of the same concept.

I am certainly not suggesting that initial intuitive means should be avoided in the teaching of science and mathematics.

Rather, my suggestion is that the teacher has to start as early as possible refining and both gradually and systematically generalizing the models on which his teaching is based.

That recommendation is less trivial than it may sound.

Considering that until the age of 11—12 the child is in the concrete operational period, one is inclined to resort usually to a large variety of concrete instructional materials and models. The intuitive interpretations created in this way tend to become very rigid and to obstruct the further process of abstraction and generalization. Our suggestion is to start as early as possible, during the concrete operational period, including didactical requirements and forms of activity which would enable the child to gradually assimilate concepts of higher complexity and abstraction.

For instance, the number concept initially understood as representing collections (sets) of objects should be related as early as possible to the concept of measure which would prepare the understanding of rational and irrational numbers. In the same general context one has to prepare the introduction of multiplicative structures (multiplication, division, proportions) *even before the formal operational period* in order to avoid the rigid association of multiplication with repeated addition and of division with practical fragmentation.

I do not argue that one has to force upon the child concepts which he is not yet intellectually mature enough to understand. What I am advocating is that during the concrete operational period, when the basic notions of arithmetic and science are introduced, one has to find methods for preparing intellectual progress towards higher order concepts. This can be done while still using concrete behavioral procedures.

In short, considering the resistance to alteration of initial intuitive crystalizations in the child, one has to lay the foundations as early as possible, by adequate didactical means, for the subsequent refinement and generalization of the respective conceptions.

CHAPTER 18

SUMMARY AND DIDACTICAL IMPLICATIONS

THE ROLE OF INTUITION: A SUMMARY

The term intuition refers to a large variety of cognitive phenomena. To some authors, intuition means the fundamental source of certain knowledge. For others, intuition represents a particular method for grasping the truth, the essence of reality. In a third usage an intuition is a special type of cognition characterized by self-evidence and immediacy. In our interpretation the term is mainly related to this third meaning. An intuition is a cognition characterized by the following properties:

Self-evidence and immediacy. An intuitive cognition appears subjectively to the individual as directly acceptable, without the need for an extrinsic justification — a formal proof or empirical support.

Intrinsic certainty. Although self-evidence and certainty are highly correlated they are not reducible one to the other. The feeling of certainty may have an extrinsic source (the authority of the teacher, the support of a proof etc.). Experimental findings have shown that, in certain circumstances, a statement which appears obvious to the subject is, nevertheless, accepted with some doubts. High intuitiveness implies the combination of a strong feeling of evidence with a high level of confidence. For this reason, we have suggested the use of the formula $I = \sqrt{C \times O}$ for measuring intuitiveness in which C stands for confidence and O for obviousness.

Perseverance. Intuitions are stable acquisitions, resistant to alternative interpretations.

Coerciveness. Intuitions exert a coercive effect on the individual's reasoning strategies and on his selection of hypotheses and solutions. In the history of science and mathematics the coercive influence of intuitions has often determined the perseverance of erroneous interpretations. Similar situations take place during the individual's intellectual development. Immature, erroneous cognitive attitudes may survive in the individual even after he has been provided with adequate representations and solutions.

Theory status. An intuition is a theory, never a mere skill or perception. It expresses a general property perceived through a particular experience.

Extrapolativeness. It is through intuition that we extrapolate indirectly from a limited amount of information to data which are beyond our direct grasp (for instance from the finite to the infinite). One may assume on the basis of research evidence that the individual is intuitively able to attain by extrapolation the notion of a potentially infinite sequence but not that of an actually infinite set.

Globality. An intuition is a structured cognition which offers an unitary global view, in contrast to logical thinking which is explicit, analytical and discursive. Intuitions attain globality through a selection process which tends to eliminate the discordant clues and to organize the others in conformity with a unitary, compact meaning.

Implicitness. Although apparently self-evident, intuitions are in fact based on complex mechanisms of selection, globalisation and inference. But this activity is generally unconscious and the individual is aware only of the final product, the apparently self-evident, intrinsically consistent cognitions. The tacit character of intuitive elaborations explains the difficulty of controlling and influencing them.

The cognitive-behavioral function of intuitions. In our opinion intuition is the analog of perception at the symbolic level. It has the same behavioral task as perception, namely to prepare and to guide our mental or practical activity. Therefore an intuitive conception must possess a number of features analogous to that of perception: globality, structurality, imperativeness, direct evidence, a high level of intrinsic credibility. In this way intuitions are able to inspire and guide our intellectual endeavors firmly and promptly even in a situation of uncertain or incomplete information.

The survival of intuitive components in scientific reasoning — historically and individually — may then be explained by that profound necessity of human beings to rely in their reasoning activity upon apparently certain, evident, trustworthy conceptions. The appearance of direct trustfulness is created automatically by a number of special mechanisms.

THE CLASSIFICATION OF INTUITIONS

According to a first classification, which considers the relation between intuitions and solutions, one may distinguish affirmatory, conceptual, anticipatory and conclusive intuitions.

Affirmatory intuitions are representations or interpretations of various facts accepted as certain, self-evident and self-consistent. An affirmatory intuition may refer to the meaning of a concept (for instance the intuitive meaning of notions like force, energy, point, line etc.); to the meaning of a relationship or a statement (for instance a force is necessary in order to maintain the motion of a body); to the acceptance of an inference, which may be either inductive or deductive (for instance from $A = B$ and $B = C$ one deduces, as intuitively evident, that $A = C$). Affirmatory intuitions may be further classified into ground (common, basic), or individual, intuitions.

Conjectural intuitions are assumptions associated with the feeling of certainty. For instance "I am sure that you will become an excellent engineer". Such intuitions play an important role in the diagnostic capacity of professionals.

Anticipatory intuitions are also conjectures but they have been classified separately since they belong explicitly to a problem-solving activity. An anticipatory intuition is the preliminary, global view of a solution to a problem, which precedes the analytical, fully developed solution. Not every hypothesis is an intuition; only those hypotheses which are associated, from the beginning, with the feelings of certainty and evidence, are anticipatory intuitions. The contradictory nature of anticipatory intuitions (and of intuitions in general) is expressed in the introspective revelations of scientists and mathematicians. In its initial form the solution is perceived simultaneously as certain and imperative, yet also "tenuous" and "transient".

Conclusive intuitions summarize in a conclusive, global vision the essential ideas of the solution of a problem previously elaborated. This full, global view adds to the formal, analytical construction a feeling of an intrinsic, direct certitude.

A second basic classification refers to *primary* and *secondary intuitions*.

Primary intuitions develop in individuals independently of any systematic instruction as an effect of their personal experience. They may be either pre-operational or operational; this distinction parallels that of Piaget with regard to the developmental stages. Pre-operational intuitions are based on configurations while operational intuitions are based on operational structures (for instance acceptance of various types of conservation as *a priori* evident, the intuitive understanding of mechanical causality). Operational intuitions which develop during the concrete operational period remain as stable acquisitions for the whole of one's life. During the formal operational period no fundamental changes take place spontaneously at the intuitive level. As an effect of the development of formal capacities the intuitive acquisitions may gain in precision and clarity but they remain basically the same. A pupil who learns about the nature of rational and real numbers may get a neater and clearer understanding of the concept of number, but the intuition of number will remain naturally attached to the ideas of cardinality and order which characterize the natural numbers.

We assume that under a systematic, instructional influence new intuitions, new cognitive beliefs may be created; these we term *secondary intuitions*. Such a process implies, in our view, the personal involvement of the learner in an activity in which the respective cognitions play the role of necessary, anticipatory and, afterwards, confirmed representations. One may learn about irrational numbers without getting a deep intuitive insight of what the concept of irrational number represents. Only through a practical activity of measuring may one discover the meaning of incommensurability and the role and meaning of irrational numbers.

INTUITIONS AND MODELS

We say that a system B represents a model of a system A if, on the basis of a certain isomorphism, a description or a solution produced in terms of A may be reflected consistently in terms of *B* and *vice-versa.* If a notion is not representable intuitively one tends to produce a model which can replace the notion in the reasoning process. We are referring here especially to substitutes which are able to translate the concept in sensorial behavioral terms. These are *intuitive models.*

Models are produced either deliberately or automatically. Very often, in a reasoning process, the search and solution strategies are influenced by models functioning tacitly, which are then beyond direct conscious control. Such tacit, automatically produced models frequently determine the construction of intuitive structures.

Various types of intuitive models may be described. If the original and the model belong to two different systems we have an *analogy* (for instance an electrical current and a flow of liquid). In the case of a paradigmatic model the original is a certain category of objects or phenomena, while the model is provided by an example or a sub-class of the category considered (for instance, for many children water is a model for defining and identifying liquids). If a certain phenomenon, with intuitive qualities, may help the understanding of a more complex related phenomenon, then we have a phenomenological primitive (the behavior of a spring, compressed and tending to expand back, may confer an intuitive meaning to various phenomena in which conservation and transformation of energy intervene). In the case of diagrammatic models there are various types of graphical representations which play the role of models. Diagrammatical models are of fundamental importance in translating abstract relationships into intuitive representations (for instance graphs of functions, tree diagrams, Venn diagrams etc.). They may be didactically useful only if the student has learned the proper syntax of the model and the laws of its correspondence with the original.

Models must have a number of features in order to be really useful as heuristic devices. The model must present a high degree of natural, consistent and structural correspondence with the original. It must also correspond to human information processing characteristics (spatial, visual representability, behavioral manipulability, finiteness etc.). Thirdly, the model must enjoy a relative autonomy with respect to the original. This implies that the model must be internally coherent and structured, governed by its own laws. A model which is dependent at every step on the dictates of the original is heuristically useless.

Very often, the child faces difficulties in his learning, understanding and solving endeavors because his reasoning strategies are controlled by inadequate tacit models. The teacher should try to identify such models and to help the student to become aware about them. He has to help the student to correct his mental models or to resort to more adequate ones if this is the case. The intellectual progress of the child is not reducible to a formal conceptual development. Intuitions and intuitive models represent powerful components of any productive mental activity, the impact of which has to be taken into account in the instructional process.

THE MECHANISMS OF INTUITIONS

Intuitions are generated by experience, i.e. by practical situations in which the individual is systematically involved and which require anticipatory, global, well structured representations and evaluations. An intuition may then be described as a well stabilized cluster of expectations with respect to certain situations.

Various mechanisms have been identified which participate in the process of generating intuitions.

Overconfidence. In order to stabilize an intuitive cognition so that it may be acted upon firmly and promptly in uncertain situations the individual usually has to overestimate its probability of being correct. Overconfidence is achieved by distorting the importance of various pieces of information. The apparently favorable ones for the accepted solution are preferred while those which contradict it are minimized or simply overlooked. This selection process is automatic (in the case of intuition).

Dramatization. In order to increase the plausibility of an intuitive interpretation, the individual tends, automatically, to dramatize the sequences of facts i.e. to combine them so as to present a logically acceptable coherence.

Premature closure. One tends automatically to stop the search for new information and the examination of arguments sooner than would be objectively justified. One thus gets an intuitive view, a prematurely closed representation which has the appearance of certitude and intrinsic consistency.

The primacy effect. It has been shown that the individual often tends to favor a certain interpretation or solution only because it was the first to be chosen and accepted. One tends to avoid alternative interpretations which would upset the attained equilibrium. Knowledge tends to remain structured, apparently coherent, and this contributes to the "epistemic freezing" which characterizes intuitions.

Factors of immediacy. Intuitions are immediate, apparently self-evident cognitions. Various factors are automatically elicited to contribute to the effect of immediacy:

— Visualization: A visual image delivers simultaneously and in a relatively structured manner most of the information related to a situation (this is not the case, for instance, with acoustic messages). In addition, a visual image is more personal and more behaviorally involved than a conceptual structure.
— Availability. A solution may be accepted as certain not because it objectively fits formal requirements but because it is easily *accessible.* If a certain element in a class can be more easily detected than others, one tends to consider it also as being more frequent.

- Anchoring. A certain salient feature may become decisive in the individual's intuitive interpretation not because it is objectively decisive but merely because it is more salient.
- Representativeness. The probability of an uncertain event or sample belonging to a class may be determined not by objective considerations but by its superficial, apparent representativeness of the respective category. The sequence ABABB is considered intuitively to be more representative for a random process than, for instance, the sequence AAAAA, and therefore more likely to appear.

Factors of globality. An intuition takes the form of a global, synthetic view but this is the effect of some integrative tacit procedure. One is that referred to by Gestalt psychology — the law of closure: a group of elements tend to organize themselves so as to constitute a "Gestalt", an ensemble defined by a unitary significance. Imagination adds the apparently absent fragments in order to complete the whole.

It has also been shown that the process of interpretation is often achieved in conformity with some tacit algebraic rules. In young, pre-operational children these are additive rules while in older children the integration is based mainly on multiplication rules. (Areas of rectangles, for instance, are intuitively estimated and compared either by adding or by multiplying tacitly the dimensions.) Pre-operational children tend also to integrate, through extrapolation, from one salient feature (for instance from one dimension to the whole object). When a principle common to a variety of events or situations is discovered, it contributes to the intuitive grasp of the essence of the respective phenomena.

CONFLICTS AND COMPROMISES

Certain circumstances may generate contradictory intuitions. (Two segments of different lengths are intuitively supposed to contain the same — infinite — number of points. But, from a figural perspective, they are supposed, also intuitively, to contain a different number of points.)

Conflicts appear also between intuitive interpretations and formal ones (acquired by instruction). In children, such contradictory interpretations may coexist. But very often the intuitive representation is stronger and tends to annihilate the formal conception. Pupils easily forget that weight presupposes a gravitational force, that a launched body maintains its state of uniform motion if no force intervenes etc. Sometimes, two contradictory conceptions may merge into a new compromise intuition. For example, the dynamic representation of infinity is a compromise between the finite structure of intellectual schemas and the formally acquired ideas of infinity.

It is recommended that the student should be made aware of his tacit mental conflicts in order to strengthen the control of the taught conceptual structures over the primary intuitive ones.

Intuitions represent powerful, coercive forces in mental activity. They act, sometimes, in an overt manner but more often implicitly. They may represent a source of important productive ideas but frequently they distort or hinder the individual's mental strategies.

The educational problem is not to eliminate intuitive representations and interpretations. This, in our view, would be impossible and certainly not desirable. Rather, the educational problem is to develop the capacity of the student to analyse and keep under control his intuitive conceptions and to build new intuitions consistent with formal scientific requirements.

DIDACTICAL IMPLICATIONS

For a very long time, reasoning has been studied mainly in terms of propositional networks governed by logical rules. Consequently, the instructional process, especially in science and in mathematics, has tended to provide the learner with a certain amount of information (principles, laws, theorems, formula) and to develop methods of formal reasoning adapted to the respective domains.

What has been shown in this work is that, beyond the dynamics of the conceptual network, there is a world of stabilized expectations and beliefs which deeply influence the reception and the use of mathematical and scientific knowledge. For the science teacher and the teacher of mathematics it is of fundamental importance to identify these intuitive forces and to take them into account in the instructional process.

Let me emphasize some aspects:

(1) Mathematical entities such as numbers and geometrical figures do not have an external, independent existence as the objects of the empirical sciences do. In mathematics, we deal with entities whose properties are completely fixed by axioms and definitions. Dealing with such entities requires a mental attitude which is fundamentally different from that required by empirical, materially existing realities. When one defines a category of concrete objects one is aware of the fact that the definition only approximates full knowledge of the respective category. The definition of solids, for instance, only approximates the properties of solids. In order to know more about solids one has to study solids, not analyze definitions. New properties, not deducible from the definition, may be discovered. With mathematical concepts things are different. Mathematical entities owe their entire existence, all their properties, to what has been imposed formally. This creates a new didactical situation; the student has to learn to understand and use the mathematical concepts *in absolute conformity with the corresponding axioms and definitions*. This is an important but very difficult didactical task. Mathematical concepts possess, generally, an intuitive loading, as every concept does. Pictorial representations, various types of interpretative or

solving models such as paradigms and analogies, confer on mathematical concepts a certain kind of direct accessibility which is required by the dynamics of productive reasoning. Our mathematical thinking remains profoundly rooted in our adaptive practical behavior, which implies spatial representability, concrete consistency, fluent continuity. *The main problem is to learn to live with the intuitive loading of concepts — necessary to the productive fluency of reasoning — and, at the same time, to control the impact on the very course of reasoning of these intuitive influences. For this, the student has to learn to become aware of the exact, formal meaning and the implications of the mathematical concepts, on the one hand, and the underlying intuitions on the other.*

It is difficult for a pupil to accept that a square is a parallelogram, because the notion of parallelogram implies, intuitively, unequal sides and angles (as a consequence of the usual paradigmatic model). Formally, a square *is* a parallelogram because the definition of a square implies all the properties of parallelograms.

The definition of a tangent does *not* imply a single contact point with a curve but only the property of expressing the slope of a curve at a certain point. The intuitive loading includes the uniqueness of the point (the paradigmatic model of the circle).

Through metacognitive techniques the student has then to learn to see clearly the formal properties of the mathematical concepts used and to understand the intuitive sources of his misconceptions.

In the empirical sciences it is also important to possess clear, explicit definitions of the concept used. But in this case, it is not the definition which imposes the properties of the concept. On the contrary, it is the complex reality of the respective phenomenon which remains the permanent source for enriching and rendering more precisely the notion considered. The mental attitude is different. The student has to learn to cope in his productive reasoning with these two different types of situations: in mathematics it is the formal, axiomatic basis which is decisive; in the empirical sciences it is the experimental evidence which decides. It may not be difficult to accept the distinction theoretically, but it is very hard to assimilate it practically, intuitively and operationally because *intuitively* the distinction does not exist.

(2) A second aspect, related to the first, is the fact that the student has to learn to analyze and formalize his primary intuitive acquisitions. This implies learning how to abstract formal structures from practical realities and intuitive interpretations and how to describe them explicitly.

We know intuitively what a circle is, or a square, or a triangle. We are able to recognize these figures on the basis of tacit considerations. Let us try to describe the common, general properties in a precise, complete manner. Such exercises represent the complement of those first mentioned (starting from the definition). The fundamental task of these two types of activities is that of training the student to become able to control his primary intuitive acquisitions productively, in a way which keeps alive their creative contribution without the student being misled by them.

(3) Possible conflicts between the formal and the intuitive interpretation of concepts and operations have been mentioned.

An essential recommendation is to create didactical situations which can help the student to become aware of such conflicts. However, rendering manifest the latent conflicts does not solve the problem by itself. This procedure has to be associated with the already mentioned activity of analyzing explicitly the properties — as stated by definitions — of the mathematical entities considered (in contrast with the intuitive interpretations).

For instance, if one compares the sets of points in two line segments of different lengths, one obtains two opposite conclusions, both intuitively valid: the two sets are equivalent because both are infinite; The two sets are not equivalent because the longer segment contains more points. The role of making the student aware of the conflict is to help him to understand intuitively that in the domain of infinity, intuitive arguments may be misleading. By analyzing the concept of a point one emphasizes that, formally, a point has no dimensions, no real content. The longer segment has more points if the points are considered in their intuitive interpretation (small spots). But since mathematical points have no dimensions, it is only through formal considerations that the problem can be resolved.

Intuitively, multiplication "makes bigger". But when multiplying a number with a decimal smaller than one this intuitive rule does not hold any more. One then gets a conflict (a multiplication which "makes smaller"). In order to overcome it one has not only to become aware of the contradiction. One has also to analyze the concept of multiplication, to understand that repeated addition, the source of the conflict, is only a particular model, and to realize the more general meaning of multiplicative structures with their formal laws.

(4) Generally, when introducing new mathematical or scientific concepts, especially during the first school years, one uses intuitive models. Such intuitive interpretations correspond to the properties of the concrete operational period and render the notions more accessible to the students. Numbers are taught as expressing cardinals of sets, arithmetical operations are related to practical activities, geometrical notions are based on concrete spatial properties, etc. It is supposed that later on more general, abstract meanings and definitions will be introduced. But the primacy effect suggests, and research has confirmed, that these initial intuitive interpretations become very strongly attached to the respective concepts and, consequently, it becomes very difficult to escape from their impact. Yet it is impossible to avoid using intuitive means initially when introducing new mathematical concepts. While there is no general recipe for solving the dilemma, this is what we recommend. One has to start, as early as possible, preparing the child for understanding the formal meaning and the formal content of the concepts taught. This may be done, first of all, by revealing the relationships

between concepts and operations and by rendering explicit the underlying common structures of different concepts and operations. Multiplication and division, for instance, are inverse aspects of multiplicative structures which, in turn, are related to proportional reasoning. Addition and subtraction are intuitively based on opposite practical operations. The algorithms are different but mathematically the two operations are deeply related. In fact, problems which include the notion of addition are sometimes, solved by subtraction and vice versa. By getting used to such situations the student learns to detach the mathematical operation from a certain particular, intuitive model and to see these operations in their general, formal context.

(5) It is well known that one of the difficulties in teaching the use of proofs in mathematics is the fact that, very often, proofs appear to the student to be useless. Many mathematical statements seem to be self-evident and therefore a proof is considered to be superfluous. In such cases intuitions help the student to accept the statement but prevent him from accepting the necessity of the proof.

One of the fundamental tasks of mathematical education — as has been frequently emphasized in the present work — is to develop in students the capacity to distinguish between intuitive feelings, intuitive beliefs and formally supported convictions. In mathematics, the formal proof is decisive and one always has to resort to it because intuitions may be misleading. This is an idea which the student has to accept theoretically but that he has also to learn to practise consistently in his mathematical reasoning.

On the other hand, it would be a serious mistake to undermine the students' confidence in their intuitions. In order to avoid this, it is important to develop in students the conviction that: (a) one possesses also correct, useful intuitions and (b) that we may become able to control our intuitions by assimilating adequate formal structures.

A similar situation appears also in empirical sciences. As has been seen, very often our intuitions in science may be misleading. The student has to be aware of this and, at the same time, he has to learn that experimental evidence and formal analyses of concepts and statements are means by which one achieves objective validation of scientific truths.

Certainly, the student has to realize the fundamental difference between a formal proof in mathematics and an experimental confirmation. A formal proof guarantees the universal validity of a statement, while additional empirical evidence only increases the probability of the respective assertion. As we have already emphasized, the distinction is not intuitively evident and special care has to be taken in mathematics and science education in order to develop in students the understanding of that idea.

(6) When referring to the development of intuitions one has to consider also anticipatory intuitions. Though mathematics is a deductive system of knowledge, the creative activity in mathematics is a constructive process in which inductive procedures, analogies and plausible guesses play a funda-

mental role. The effect is very often crystallized in anticipatory intuitions. Much more attention should be given, in our opinion, to educating in students sensibility for similarities, the ability to identify isomorphisms and describe common structures. As Poincaré and Polya have shown, new fruitful ideas are frequently suggested by analogies between apparently very different mathematical entities.

On the other hand, the student should learn to evaluate the plausibility of preliminary solutions. This cannot be done, in our opinion, by mere verbal explanations. It is rather a problem of practical training in the course of which systematic discussions in the classroom, about competing hypotheses, have to be encouraged. But we do not consider that, when conjecturing during the search for a solution, one has to evaluate explicitly the chances of every anticipatory intuition being correct. This is impossible for two reasons. One is that the solving process is chiefly an automatic one. We cannot control explicitly every step of our solving endeavor. The second reason is that anticipatory intuitions appear to the individual as certain truths, not as mere conjectures.

When referring to the evaluation of the plausibility of anticipatory intuitions we are mainly considering a kind of automatic, tacit selection of hypotheses according to their plausibility. In our opinion it may be possible to develop such tacit selection processes by a systematic activity of discussing in the classroom competing interpretations and anticipatory global solutions of problems.

Anticipatory intuitions do not appear spontaneously ex nihilo. They are influenced by the lines of force determined by tacit intuitive tendencies which may not be in conformity with the formal conceptual constraints. This means that an important role in educating the solving capacity of the student is played by the development of his correct, intuitive interpretations and by his capacity to control his intuitions conceptually.

Anticipatory intuitions are conjectures associated with a feeling of total confidence. This raises again the problem already discussed above. If the student finds that his conjectures may be misleading to such a high degree, he may not be willing to make any more conjectures (at all) or at least to express them publicly (in the classroom). Such an effect would certainly block the student's solving capacity. The student has then to learn *to accept the risk* of erroneous guesses (even publicly). He should understand that this is the way in which *everybody* solves problems — not only the novice. Certainly we do not consider wild guesses, but only plausible conjectures based on serious preliminary analyses. On the other hand, one has to develop the student's capacity to analyze and check his findings, his anticipatory solutions, both formally and intuitively. Our belief is that it is possible to develop, through adequate training, the student's intuitive feelings of incongruences, of incompleteness of arguments, of flaws in lines of thought. This, together with the capacity to analyze formally and systematically the preliminary solution (the anticipatory intuition) represents an essential condition of success in a problem solving endeavor.

(7) Should one encourage the use of intuitive means in science and mathematics education?

As we have emphasized several times, intuitive interpretations and solutions are a *sine-qua-non* component of every productive reasoning endeavor. The educational problem is not to eliminate intuitions — affirmatory or anticipatory. This in our view is impossible. The educational problem is to develop new, adequate, intuitive interpretations as far as possible, together with developing the formal structures of logical reasoning.

This may be done especially through appropriate practical activities and not through mere verbal explanations. Intuitions are in our opinion by their function and their nature behaviorally, practically orientated.

However, not everything in mathematics or science lends itself to an intuitive interpretation. This has to be clearly understood by the students. There are scientific and mathematical concepts and statements which are mere formal constructs beyond any possible intuitive representation. This is so in mathematics because mathematics is by its very nature a formal, deductive system of knowledge. It is so in empirical sciences mainly because in certain domains one deals with phenomena with no direct sensorial meaning (for instance, elementary particles).

Our claim is that one should never use artificial intuitive strategies when in fact no adequate intuitive interpretation is possible. On the contrary, one should, in our opinion, use such opportunities in order to make the student realize intuitively the fundamental role of logical constructs in both mathematics and empirical sciences.

CONCLUDING REMARKS

Mathematical and scientific reasoning in general are not reducible to formal conceptual structures. The history of mathematical and scientific acquisitions has been influenced by the profound tendency of individuals to produce tacitly mental devices which enable them to believe directly in the objective validity of their conceptions, even before a complete justification is reached. We consider that the emergence of apparently self-evident, self-consistent cognitions — generally termed intuitions — is a fundamental condition of a normal, fluent, productive reasoning activity. An intuition is a complex cognitive structure the role of which is to organize the available information (even incomplete) into apparently coherent, internally consistent, self-evident, practically meaningful representations.

But mathematics is by its very nature a formal, axiomatically organized system of knowledge. Every statement in mathematics has to be accepted only on the grounds of an explicit, complete proof (apart from a few axioms and primitive notions).

We get, then, a profound contradiction between *the nature of mathematics* and the nature of *mathematical reasoning*. The dynamics of mathematical reasoning — and, generally, of every kind of scientific reasoning — include various psychological components like beliefs and expectations, pictorial prompts, analogies and paradigms. These are not mere residuals of more primitive forms of reasoning. They are genuinely productive, active ingredients of every type of reasoning.

This leads — as I have said — to a profound, dialectic contradiction. One needs intuitive prompts and incentives in order to be able to think productively. But at the same time, the main endeavor of mathematical reasoning is to become "purified" by eliminating all the extra-logical supports and arguments.

It is highly illuminating to compare the obstacles, difficulties and distortions which have appeared in the history of mathematics with those which emerge during childhood and in the instructional process. Basically, the same types of conflicts may be identified. Intuitive factors — the quest for practicality, for behavioral interpretations, for visual, spatially consistent expressions — have profoundly influenced the historical development of the number concept, of the various geometries, of the infinitesimal calculus, of the concept of infinity, etc. Similar phenomena may be detected during the instructional process. This supports the hypothesis that intuitive forms of reasoning are not only a transitory stage in the development of intelligence. On the contrary, typical intuitive constraints influence our ways of solving and interpreting at every age. Even when dealing with highly abstract concepts, one tends to represent them almost automatically in a way which would render them intuitively accessible. We tend automatically to resort to behavioral and pictorial representations which can confer on abstract concepts the kind of manipulatory features to which our reasoning is naturally adapted. It has been proved that even long after the student has acquired the adequate, highly abstract knowledge referring to a certain mathematical notion, the primitive, intuitive model on which this notion was originally built may continue to influence, tacitly, its use and interpretation. It is difficult, for instance, not to represent multiplication in terms of repeated addition, or division as material fragmentation. It is difficult not to interpret the concept of limit in terms of an actual, dynamic process, or a point and a line in terms of pictorial representations.

Certainly one has to distinguish between the capacity to define explicitly a concept or an operation and the active intervention of a certain concept or operation in a practical solving endeavor. The same person who is able to define a concept correctly may use the same concept incorrectly when solving a certain problem under the influence of intuitive, tacit constraints. Knowing formally the definition of the altitude of a triangle does not imply that the student will recognize or draw correctly the altitude in any triangle no matter what its shape or position.

Conflicts between intuitive and scientific interpretations appear also in empirical sciences. However, there is a basic difference. In mathematics, the main conflict is between the abstract, purely formal nature of mathematical concepts and various intuitive tendencies of the reasoning activity. In empirical sciences the basic conflict is between our intuitions based on limited terrestrial experience and the logical structure of sciences based on data and considerations which largely transcend our daily experience.

Mathematical and science education cannot ignore the impact of intuitive forces on the student's ways of reasoning. While much can be learned from history, recent studies have made an important contribution in this domain. We know now that intuitive mechanisms are organized in firm, coherent complex structures very resistant to alterations. Much more experimental evidence is needed, but the teacher himself may very often discern such elements of resistance which conflict with the taught concepts.

One has to bear in mind that intuitively based conceptions cannot be eliminated simply by mere verbal explanations. *Intuitions are always the product of personal experience, of the personal involvement of the individual in a certain practical or theoretical activity.* Only when striving to cope actively with certain situations does one need such global, anticipatory, apparently self-consistent representations. The development of new — secondary — mathematical and scientific intuitions implies, then, didactical situations in which the student is asked to evaluate, to conjecture, to predict, to devise and check solutions. In order to develop new, correct probabilistic intuitions, for instance, it is necessary to create situations in which the student has to cope, *practically*, with uncertain events.

It is important to emphasize that new, correct intuitions do not simply replace primitive, incorrect ones. Primary intuitions are usually so resistant that they may coexist with new,.superior, scientifically acceptable ones. That situation very often generates inconsistencies in the student's reactions depending on the nature of the problem. A student may understand logically and intuitively that when tossing a coin several times, each outcome has the same probability. Nevertheless he may still feel intuitively, that, after getting "tails" 3—4 times in succession, there is a greater likelihood of getting "heads" on the next toss.

In mathematics education the conflicting nature of mathematical representations (if considered psychologically) has given rise to two opposite didactical strategies. On one hand, many curricula and text-book writers have tended to emphasize the intuitive, pictorial components, apparently in order to meet the child's strong need for intuitive representations. Text-books became, then, full of beautifully colored images and diagrams. On the other hand, especially during the sixties and seventies, other authors, mainly under the influence of the Bourbaki group, tried to set up programmes and textbooks in which the body of knowledge was presented axiomatically. In our opinion both strategies were mistaken because each of them considered only a half of the complex structure of mathematical concepts which, psychologically, are both intuitively and formally based.

By exaggerating the role of intuitive prompts, one runs the risk of hiding the genuine mathematical content instead of revealing it. By resorting too early to a "purified", strictly deductive version of a certain mathematical domain, one runs the risk of stifling the student's personal mathematical reasoning instead of developing it.

A third attempt was that of Zoltan Dienes who tried to synthesize the two lines of thought in what has been called "structured materials". One may now affirm that the success of the Dienes approach, used without systematic preliminary investigations, was also very limited.

The whole problem is very complex and cannot be solved by some particular didactical techniques. The educational strategy in both mathematics and empirical sciences has to be built on the basis of a profound knowledge of the nature and the historical and ontogenetic development of the respective concepts.

It seems to me beyond any doubt that it is inadmissible to introduce new concepts in school programmes, especially in science and mathematics without a thorough preliminary psychological and psycho-didactical investigation.

This is probably the main lesson which may be drawn from our analysis. Too much time and energy have been wasted, too many mistakes have already been made with new inadequate projects.

In order to cope successfully with the instructional problems, one has first to have a good, serious understanding of the psychological aspects of the concepts involved. Much more research is needed in this respect. We have to know what are the tacit interpretations the student attaches to these concepts, what are the intuitive reactions, the intuitive models he produces, the impact they may have on the acquisition of the new concepts. On the other hand one has to evaluate the effect of the various didactical means on the complex and labile relationships between the intuitive loading of the concepts taught and their formal structure. An inadequate strategy may destroy the productive interaction of these two components.

A joint effort of psychologists, mathematicians and scientists, of teachers and researchers, is necessary in order to produce new, adequate, more efficient programmes and didactical solutions in science and mathematics education.

BIBLIOGRAPHY

Adi, H., Karplus, R. and Lawson, A. E. (1980), 'Conditional logic abilities on the four-card problem: assessment of behavioral and reasoning performances', *Educational Studies in Mathematics* **11**(4), 479—496.

Anderson, H. (1979), 'Algebraic rules in psychological measurement', *American Scientist* **67** (5, September—October), 555—563.

Bachelard, G. (1980), *La formation de l'esprit scientifique*, Librairie Philosophique, J. Vrin, Paris.

Bar-Hillel M. (1982), 'Studies of representativeness', in Kahneman D., Slovic P. and Tversky, A. (eds.), *Judgement Under Uncertainty: Heuristics and Biases*, 69—83, Cambridge University Press, Cambridge.

Barrat, B. (1975), 'Training and transfer in combinatorial problem solving: The development of formal reasoning during early adolescence', *Developmental Psychology* **11**(6), 700—704.

Bartlett, F. (1958), *Thinking*, London: George Allen and Unwin.

Bell, A. W. (1976), 'A study of pupils' proof-explanations in mathematical situations', *Educational Studies in Mathematics* **7**, 23—40.

Bereiter, C., Hidi, S. and Dimitroff, G. (1979), '*Qualitative changes in verbal reasoning during middle and late childhood*', *Child Development* **50**, 142—151.

Bergson, H. (1954), *Creative Evolution*, Translated by A. Mitchell. Macmillan & Co. Ltd., London 1954, First edition 1911.

Beth, E. W. and Piaget, J. (1966), *Mathematical Epistemology and Psychology*, Reidel, Dordrecht.

Bishop, A. J. (1979), 'Visualising and mathematics in a pre-technological culture', *Educational Studies in Mathematics* **10**, 135—146.

Bolton, N. (1972), *The Psychology of Thinking*, Methuen, London, 1972.

Bruner, J. (1965), *The Process of Education*, Harvard University Press, Cambridge, Massachusetts.

Bunge, M. (1962), *Intuition and Science*, Prentice-Hall, Inc., Englewood Cliffs.

Campbell, N. R. (1920), *Physics, the Elements*, Cambridge Univ. Press, Cambridge.

Carroll, L. (1895), 'What the tortoise said to Achilles', *Mind* **4**, 278—80.

Cavaillès, J. (1969), *Studii asupia teoriei multimilor*, Translated by C. Popescu-Ulmu, Editura Stientifica, Bucuresti. (J. Cavaillès, Philosophie Mathématique, Hermann, Paris, 1962.)

Christensen-Szolansky, J. J. and Bushyhead, J. B. (1981), 'Physician's use of probabilistic information in real clinic setting', *Journal of Experimental Psychology: Human Perception and Performance* **7**, 928—935.

Clagett, M. (1962), *Critical Problems in the History of Science*, The University of Wisconsin Press, Madison.

Clement, J. (1982), 'Student's preconceptions in introductory mechanics', *American Journal of Physics* **50**, 66—71.

Cohen, P. J. (1966), *Set Theory and the Continuum Hypothesis*, W. A. Benjamin, Inc. New York.

Courant, R. and Robbins, H. (1969), *What is Mathematics? An Elementary Approach to Ideas and Methods*, Oxford University Press, Oxford.

Delachet, A. (1961), *L'analyse mathématique*, Que Sais-Je? series, P.U.F., Paris.

Descartes, R. (1967), *The Philosophical Works*, vol. 1, Translated by E. S. Haldane and G. R. T. Ross, The University Press, Cambridge.

DiSessa, A. (1982), 'Unlearning Aristotelian physics', *Cognitive Science* **6**, 37—75.

DiSessa, A. H. (1983a), 'Phenomenology and the evolution of intuition', in D. Gentner and A. L. Stevens (eds.), *Mental Models*, pp. 15—33, Lawrence Erlbaum Associates, London.

DiSessa, A. A. (1983b), *Intuition as Knowledge*, Unpublished.

Duval, R. (1983), L'obstacle du dedoublement des objets mathématiques, *Educational Studies in Mathematics* **14**(4), 385—414.

Echternacht, G. J. (1972), The use of confidence testing in objective tests, *Review of Educational Research* **42**, 217—236.

Einstein, A. (1976), *La théorie de la relativité restreinte et générale* (Translated by M. Solovine). Gauthier-Villars, Paris 1976.

Ennis, R., Finkelstein, M., Smith, E. and Wilson, N. (1969), 'Conditional logic in children', Ithaca, N.Y.: *Cornell Critical Thinking Readiness Project* (ERIC Document Reproduction Service No. ED 040437).

Evans, J. St. B. T. (1982), *The Psychology of Deductive Reasoning*, London: Routledge and Kegan Paul.

Feller, W. (1950, 1957), *An Introduction to Probability Theory and Its Applications*, vol. I, John Wiley & Sons, New York.

Fischbein, E. (1975), *The Intuitive Sources of Probabilistic Thinking in Children*, D. Reidel, Dordrecht: Holland.

Fischbein, E. (1977), 'Image and concept in learning mathematics', *Educational Studies in Mathematics* **8**, 153—165.

Fischbein, E. (1982), 'Intuition and proof', *For the Learning of Mathematics* **3**(2), November, 8—24.

Fischbein, E. (1983), 'Intuition and analytical thinking in Mathematics Education', *Zentralblatt für Didaktik der Mathematik* **2**, 68—74.

Fischbein, E. (1983b), 'Intuition and axiomatics in mathematical education', in *Proceedings of the Fourth International Congress on Mathematical Education*. M. Zweng, T. Green, J. Kilpatrick, H. Pollack and M. Suydam (eds.) Birkhauser, Boston, 599—602.

Fischbein, E., Barbat, I. and Minzat I. (1971), 'Primary and secondary intuitions in the introduction of probability', *Educational Studies in Mathematics* **4**, 264—280 (In French). Also Appendix 1, The intuitive sources of probabilistic thinking in children, 138—155, D. Reidel, Dordrecht, 1975.

Fischbein, E. Barbat, I. and Minzat, I. (1975), 'Syllogistic reasoning in children and adolescents', *Revue Roumaine des Sciences Sociales, Serie de Psychologie* **19** (1), 21—33.

Fischbein, E., Deri, M., Nello, M. S. and Marino, M. S. (1985), 'The role of implicit models in solving verbal problems in multiplication and division', *Journal of Research in Mathematics Education* **16**(1), 3—17.

Fischbein, E. and Gazit, A. (1984), 'Does the teaching of probability improve probabilistic intuitions?', *Educational Studies in Mathematics* **15**, 1—24.

Fischbein, E., Pampu, I. and Minzat, I. (1970), 'Effects of age and instruction on combinatory ability in children', *The British Journal of Educational Psychology* **40**(3), 261—270.

Fischbein, E., Tirosh, D. and Hess, P. (1979), 'The intuition of infinity', *Education Studies in Mathematics* **10**, 3—40.

Fischbein, E., Tirosh, D. and Melamed, U. (1981), 'Is it possible to measure the intuitive acceptance of a mathematical statement?', *Educational Studies in Mathematics* **12**, 491—512.

Fischhof, B., Slovic, P. and Lichtenstein, S. (1977), 'Knowing with certainty: The appropriateness of extreme confidence', *Journal of Experimental Psychology: Human Perception and Performance* **3**, 52—564.

Franklin, A. (1978), 'Inertia in the middle ages', *The Physics Teacher*, April, 201—208.

Freudenthal, H. (1973), *Mathematics as an Educational Task*, D. Reidel, Dordrecht.

Fuson, K., Secada, W. and Hall, J. W. (1983), 'Matching, counting and conservation of numerical equivalence', *Child Development* **54**, 91—97.

Galbraith, P. L. (1981), 'Aspects of proving: A clinical investigation of process', *Educational Studies in Mathematics* **12**, 1—28.

Galilei, G. (1960), 'De Motu' (J. E. Drabkin, trans.) in I. E. Drabkin and S. Drake (trans. and eds.) Galileo Galilei: "*On motion*" *and* "*On mechanics*", Madison: University of Wisconsin Press (originally written ca. 1590).

Galileo, G. (1962), *Dialog despre cele doua sisteme principale ale lumii* (Dialogue concerning two new sciences). Roumanian translation by R. Ottone, Editura Stiintifica Bucuresti, 1962. See also:

Galilei, G. (1967), *Dialogue Concerning the Two Chief World Systems* (S. Drake trans.), Berkeley: University of California Press (Originally published 1632).

Gentner, D. (1983), 'Structure-mapping: a theoretical framework for analogy', *Cognitive Science* **7**, 155—170.

Gentner, D. and Stevens, A. L. (eds.) (1983), *Mental Models*, Lawrence Erlbaum Associates, Hillsdale, N.J.

Gettys, C. F., Kelly, C. and Peterson, C. R. (1973), 'The best guess hypothesis in multistage inferences', *Organizational Behavior and Human Performance* **10**, 364—373.

Gick, M. L. and Holyoak, J. K. (1983), 'Schema induction and analogical transfer', *Cognitive Psychology* **15**, 1—38.

Glaeser, G. (1981), 'Epistémologie des nombres relatifs', *Recherches en Didactique des Mathematiques* **2**(3), 303—346.

Griggs, R. A. (1983), 'The role of problem content in the selection task and in the THOG problem'. *In* J. St. B. T. Evans (ed.), *Thinking and Reasoning: Psychological Approaches*, London: Routledge and Kegan Paul.

Hadamard, J. (1949), *An Essay on the Psychology of Invention in the Mathematical Field*, Princeton University Press, Princeton.

Hahn, H. (1956a), 'The crisis in intuition' in J. R. Newman (Ed.), *The World of Mathematics*, Simon and Schuster, New York, 1957—1976.

Hahn, H. (1956b), 'Infinity' in J. R. Newman (Ed.), *The world of Mathematics*, Simon and Schuster, New York, pp. 1591—1611.

Hart, K. M. (ed.) (1981), *Children's Understanding of Mathematics*, 11—16, John Murray, London.

Hershkowitz, R. and Vinner, Sh. (1982), 'Basic geometric concepts — definitions and images' in: *Proceedings of the Sixth International Conference for the Psychology of Mathematics Education*, A. Vermandel (Ed.) Universitaire Instelling, Antwerpen, 18—23.

Hesse, B. M. (1966), *Models and analogies in Science*, University of Notre Dame Press, Notre Dame, Indiana.

Hill, S. (1961), *A Study of the Logical Abilities of Children*, Doctoral dissertation, Stanford University, Stanford.

Hofstadter, D. R. (1979), *Godel, Escher, Bach*, Harvester Press.

Inhelder, B., Piaget, J. (1958), *The Growth of Logical Thinking from Childhood to Adolescence*, Routledge & Kegan Paul, London.

Janvier, C. (1981), 'Use of situations in mathematics education', *Educational Studies in Mathematics* **12**, 113—122.

Johnson-Laird, P. N. (1983), *Mental Models: Toward a Cognitive Science of Language, Inference and Consciousness*, Harvard Univ. Press, Cambridge, MA.

Kahneman, D. and Tversky, A. (1982), 'Subjective probability: A judgement of representativeness', in D. Kahneman, P. Slovic and A. Tversky (eds.), *Judgement Under Uncertainty: euristics and Biases*, 32—47, Cambridge University Press, Cambridge.

Kahneman, D. and Tversky, A. (1982), 'Intuitive prediction: Biases and corrective procedures', in D. Kahneman, P. Slovic and A. Tversky (eds.), *Judgement Under Uncertainty: Heuristics and Biases*, pp. 414—421, Cambridge University Press, Cambridge.

Kant, I. (1980), *Critique of Pure Reason* (Translated by N. K. Smith), The Macmillan Press Ltd., London.

Kaput, J. J. (1979), 'Mathematics and learning: Roots of epistemological status', in J. L. Lockhead and J. Clement (Eds.), *Cognitive Process Instruction. Research on Teaching Thinking Skills.* The Franklin Institute Press, Philadelphia, Pennsylvania.

Kaput, J. J. (in press), 'Information technology and mathematics: opening new representational windows', *Journal of Mathematical Behavior.*

Kieran, C. (1981), 'Concepts associated with the equality symbol', *Educational Studies in Mathematics* **12**, 317–326.

Kilpatrick, J. (1985), 'Reflection and recursion', *Educational Studies in Mathematics* **16**(1), 1–26.

Klein, F. (1898), *Conferences sur les mathématiques, Conference VI*, A. Hermann, Librairie Scientifique, Paris.

Kline, M. (1980), *Mathematics: The Loss of Certainty*, Oxford University Press, New York.

Knifong, J. (1974), 'Logical abilities of young children — two styles of approach', *Child Development* **45**, 78–83.

Koriat, A., Lichtenstein, S. and Fischhoff, B. (1980), 'Reasons for confidence', *Journal of Experimental Psychology, Human Learning and Memory* **6**, 107–118.

Kruglansky, A. W., Ajzen, I. (1983), 'Bias and error in human judgment', in *European Journal of Social Psychology*, Febr., 1–43.

Kruglansky, A. W. and Freund, T. (1983), 'The freezing and unfreezing of lay-inferences: Effects on impressional primacy, ethnic stereotyping, and numerical anchoring', *Journal of Experimental Social Psychology* **19**, 448–468.

Kuhn, D. (1977), Conditional reasoning in children, *Developmental Psychology* **13**, 342–353.

Kuhn, S. T. (1970), *The Structure of Scientific Revolutions*, The University of Chicago Press, Chicago.

Lakatos, I. (1976), *Proofs and Refutations*, Cambridge: Cambridge University Press.

Le Roy, E. (1960), *La pensée mathématique pure*, P.U.F., Paris.

Lichtenstein, S. and Fischhoff, B.: 1977, 'Do those who know more also know more about how much they know?', *Organizational Behavior and Human Performance* **20**, 159–183.

Lichtensein, S., Fischhoff, B. and Philips, D. L. (1982), 'Calibration of probabilities: The state of the art to 1980', in D. Kahneman, P. Slovic and A. Tversky (Eds.), *Judgement Under Uncertainty: Heuristics and Biases*, p. 306–334. Cambridge University Press, Cambridge.

Luchins, A. S. (1957), 'Experimental attempts to minimize the impact of first impressions'. In C. E. Hovland (ed.), *The Order of Presentation in Persuasion*, Yale Univ. Press, New Haven, Conn.

Mason, J. and Pimm. D. (1984), 'Generic examples: seeing the general in the particular', *Educational Studies in Mathematics* **15**, 277–289.

McCloskey, M. (1983), 'Naive theories in motion', in: D. Gentner and A. L. Stevens (eds.), *Mental Models*, N. J., Erlbaum, Hillsdale.

McCloskey, M. and Kohl, D. (1983), 'Naive Physics: The curvilinear impetus principle and its role in interaction with moving objects', *Journal of Experimental Psychology: Learning, Memory and Cognition* **9**, 146–156.

McCloskey, M., Washburn, A., and Felch, L. (1983), 'Intuitive physics: The straight-down belief and its origin', *Journal of Experimental Psychology: Learning, Memory and Cognition* **9**, 636–649.

Noddings, N. and Shore, P. J. (1984), *Awakening the Inner Eye. Intuition in Education.* Teachers College Press, New York.

O'Brien, T. C. and Shapiro, B. J. (1968), The development of logical thinking in children, *American Educational Research Journal* **15**, 531–543.

O'Brien, T. C., Shapiro, B. J. and Reali, N. C. (1971), Logical thinking-language and context, *Educational Studies in Mathematics* **4**, 201–219.

Olver, R. R. and Hornsby, J. R. (1966), *On equivalence. In* J. S. Bruner, R. R. Olver and P. M. Greenfield (eds.), *Studies in Cognitive Growth*, New York: Wiley.

Oskamp, S. (1982), 'Overconfidence in case-study judgements', in D. Kahneman, P. Slovic and A. Tversky (Eds.), *Judgement Under Uncertainty: Heuristic and Biases*, Cambridge University Press, Cambridge, 287—293.

Papert, S. L. (1980), *Mind-Storms. Children, Computers and Powerful Ideas*, Basic Books, Inc., New York.

Parsegian, V. L., Meltzer, S. A., Luchins, S. A., and Kinerson, K. S. (1968), *Introduction to Natural Science, Part One: The Physical Science*, Academic Press, New York.

Parsons, C. (1960), 'Inhelder and Piaget's "The Growth of Logical Thinking". II. A logician's viewpoint', *British Journal of Psychology* **51**, 75—84.

Pedoe, D. (1958), *The Gentle Art of Mathematics.* Harmondsworth: Penguin.

Peel, E. A. (1967), 'A method for investigating children's understanding of certain logical connectives used in binary propositional thinking', *British Journal of Mathematical and Statistical Psychology* **20**, 81—92.

Piaget, J. (1930), *The Child's Conception of Physical Causality*, Kegan Paul, London.

Piaget, J. (1967), *La psychologie de l'intelligence*, Armand Colin, Paris.

Piaget, J. (1980), *Experiments in Contradiction*, University of Chicago Press, Chicago and London.

Piaget, J. and Garcia, R. (1983), *Psychogenèse et histoire des sciences*, Flammarion, Paris.

Piaget, J. and Inhelder, B. (1941), *Le developpement des quantités chez l'enfant*, Paris, Press Universitaire de France.

Piaget, J. and Szeminska, A. (1964), *La genèse du nombre chez l'enfant*, Delachaux et Niestle-Nechatel.

Pitz, G. F. (1974), 'Subjective probability distributions for imperfectly known quantities', in L. W. Gregg (ed.), *Knowledge and Cognition*, Wiley.

Poincaré, H. (1906), *La science et l'hypothese*, Flammarion, Paris.

Poincaré, H. (1913), *The Foundations of Science* (Trans. Halsted) Science Press, New York.

Poincaré, H. L. (1920), *The Value of Science*, Dover Publications Inc. New York.

Poincaré, H. (1952), *Science and Method*, Dover Publications Inc., New York.

Polanyi, M. (1969), *Knowing and Being*, Routledge and Kegan Paul, London.

Polya, G. (1954), *Induction and Analogy in Mathematics*, Princeton University Press, Princeton, New Jersey.

Popper, K. R. (1963), *Conjectures and Refutations.* London: Routledge Kegan and Paul.

Quinton, G. and Fellows, S. B. J. (1975), '"Perceptual" strategies in the solving of three-term series problems', *British Journal of Psychology* **66**, 69—78.

Reid, C. (1970), *Hilbert*, Springer, Heidelberg.

Reuchlin, M. (1973), 'Formalisation et réalisation dans la pensée naturelle: une hypothèse', *Journal de Psychologie Normale et Pathologique* **4**, 379—408.

Roberge, J. and Paulus, D. (1971), 'Developmental patterns for children's class and conditional reasoning abilities', *Developmental Psychology* **4**, 191—200.

Rosch, E. (1978), 'Principles of categorization'. *In* E. Rosch and R. B. Lloyd (eds.), *Cognition and Categorization*, Hillsdale, N.J.: Erlbaum.

Severi, F. (1951), Intuizionismo e astratismo nella matematica contemporane, in *Atti del terzo Congresso dell UMI, Ed.* Cremonese, Perella, Roma.

Shapiro, B. and O'Brien (1970), 'Logical thinking in children, ages six through thirteen', *Child Development* **41**, 823—829.

Shepard, N. R. (1978), 'Externalization of mental images and the act of creation', in B. S. Randhawa and W. E. Coffman (eds.), *Visual Learning, Thinking and Communication.* Academic Press, London.

Sieber, J. E. (1974), 'Effects of decision importance on ability to generate warranted subjective uncertainty', *Journal of Personality and Social Psychology* **30**, 688—694.

Siegler, R. S. (1979), 'Children's thinking: The search for limits', in G. I. Whilehurst, B. I. Zimmermann (eds.), *The Functions of Language and Cognition*, Academic Press, London.

Smith, E. E. and Medin, D. L. (1981), *Categories and Concepts*, Cambridge: Harvard University Press.

Spinoza, B. (1967), *Ethics and Treatise on the Correction of the Understanding* (Translated by A. Boyle). Everyman's Library Dent: London.

Stavy, R., and Stachel, D. (1985), 'Children's conception of changes in the state of matter: from solid to liquid', *Archives de Psychology* **53**, 331—344.

Steiner, M. (1975), *Mathematical Knowledge*, Cornell University Press, London.

Suppes, P. (1966), 'The axiomatic method in high school mathematics' in 'The role of axiomatics and problem solving in mathematics', *The Conference Board of Mathematical Sciences* (Ginn and Co.) Washington, D.C.

Tall, D. (1980), 'The anatomy of a discovery in mathematics research', *For the Learning of Mathematics* **1**, 25—34.

Tall, D. (1985), 'Using computer graphics programs as generic organizers for the concept image of differentiation', *Proceedings of the Ninth International Conference for the Psychology of Mathematics Education.*

Tall, D. and Vinner, Sh. (1981), 'Concept image and concept definition in mathematics with particular reference to limits and continuity', *Educational Studies in Mathematics* **12**(2), 151—169.

Taplin, J., Staudenmayer, H. and Taddonio, J. (1974), 'Developmental changes in conditional reasoning. Linguistic or logic?', *Journal of Experimental Child Psychology* **17**, 360—373.

Taylor, E. S. (1982), 'The availability bias in social perception and interaction', in D. Kahneman, P. Slovic and A. Tversky (Eds.) in *Judgement Under Uncertainty: Heuristics and Biases*, Cambridge University Press, Cambridge 1982, p. 190—208.

Thom, R. (1971), '"Modern" mathematics: an educational and philosophic error?', *American Scientist* **59**, 695—9.

Tirosh, D. (1985), *The Intuition of Infinity and its Relevance for Mathematical Education* (In Hebrew with an English abstract). Ph.D. Thesis, The Tel Aviv University, Tel Aviv 1985.

Tversky, A. and Kahneman, D. (1982), 'Judgement under uncertainty: Heuristics and biases', in D. Kahneman, P. Slovic and A. Tversky (eds.), *Judgement Under Uncertainty: Heuristic and Biases*, 3—20, Cambridge University Press, Cambridge.

Tversky, A. and Kahneman, D. (1982), 'Availability: A heuristic for judging frequency and probability', in D. Kahneman, P. Slovic and A. Tversky (eds.), *Judgement Under Uncertainty: Heuristic and Biases*, Cambridge University Press, Cambridge, 164—178.

Tweney, R. D., Doherty, M. E. and Mynatt, C. R. (1981), *On Scientific Thinking*, New York: Columbia.

Viennot, L. (1978), 'Le raisonnement spontané en dynamique élémentaire', *Revue Française de Pédagogie* **45**, 16—24.

Vinner, Sh. (1982), 'Conflicts between definitions and intuitions — the case of the tangent' in A. Vermandel (ed.), *Proceedings of the Sixth International Conference for the Psychology of Mathematical Education*, Universitaire Instelling, Antwerpen, 24—28.

Wason, P. C. (1977), 'The theory of formal operations: a critique'. *In* B. Geber (ed.), *Piaget and Knowing*, London: Routledge and Kegan Paul.

Wason, P. C. (1983), 'Realism and rationality in the selection task'. *In* J. St. B. T. Evans (ed.), *Thinking and Reasoning: Psychological Approaches*, London: Routledge and Kegan Paul.

Wason, P. and Evans, J. (1974), 'Dual processes in reasoning', *Cognition*, 1974—5, **3**(2), 141—154.

Wason, P. and Johnson-Laird, P. (1972), *Psychology of Reasoning*, Cambridge, Harvard University Press.

Wertheimer, M. (1961), *Productive Thinking* (M. Wertheimer, ed.) Tavistock Publications.

Westcott, M. R. (1968), *Towards a Contemporary Psychology of Intuition*, Holt, Rinehart and Winston, New York.

Wild, K. W. (1938), *Intuition*, Cambridge, At the University Press.

Wilder, L. R. (1965), *Introduction to the Foundations of Mathematics*, John Wiley and Sons, Inc. New York.

Wilkening, F. (1980), 'Development of dimensional integration in children's perceptual judgment: Experiments with area, volume and velocity', in F. Wilkening, J. Becker, and T. Trabaso (eds.), *Information Integration by Children*, Lawrence Erlbaum Associates, Hillsdale, New Jersey.

Wilkening, F. and Anderson, H. N. (1982), 'Comparison of two rule-assessment methodologies for studying cognitive development and knowledge structure', *Psychological Bulletin* **92**(1), 215—237.

Wilkening, F. and Anderson, N. (1984), 'Representation and diagnosis of children's knowledge structures', in N. H. Anderson (ed.), *Contributions to information integration theory*, in press.

Wittman, E. (1981), 'The complementary roles of intuitive and reflective thinking in mathematics teaching', in: *Educational Studies in Mathematics* **12**(3), 389—397.

INDEX OF NAMES

MATHEMATICS EDUCATION LIBRARY

Managing Editor

A. J. BISHOP, *Cambridge, U.K.*

Hans Freudenthal, *Didactial Phenomenology of Mathematical Structures.* x + 595 pp., 1983.
B. Christiansen, A. G. Howson, and M. Otte (eds.),, *Perspectives on Mathematics Education.* xii + 371 pp., 1986.
Adrian Treffers, *Three Dimensions. A Model of Goal and Theory Description in Mathematics Instruction – The Wiskobas Project.* xvi + 351 pp., 1987.
Stieg Mellin-Olsen, *The Politics of Mathematics Education.* xvi + 249 pp., 1987.
Efraim Fischbein, *Intuition in Science and Mathematics.* xiii + 225 pp., 1987.

GPSR Compliance
The European Union's (EU) General Product Safety Regulation (GPSR) is a set of rules that requires consumer products to be safe and our obligations to ensure this.

If you have any concerns about our products, you can contact us on

ProductSafety@springernature.com

In case Publisher is established outside the EU, the EU authorized representative is:

Springer Nature Customer Service Center GmbH
Europaplatz 3
69115 Heidelberg, Germany

www.ingramcontent.com/pod-product-compliance
Ingram Content Group UK Ltd.
Pitfield, Milton Keynes, MK11 3LW, UK
UKHW021103170926
15229UKWH00007B/31
* 9 7 8 9 4 0 1 7 3 7 3 7 1 *